Praise for *Flashes of Creation*

"Cosmology today is an established and exciting science, but in the mid-twentieth century it was looked at as somewhat disreputable. This engrossing book tells the story of the two audacious physicists who had the courage to envision the universe as a whole, disagreeing with each other but shaping our modern view of the cosmos."

—SEAN CARROLL, author of *Something Deeply Hidden*

"Paul Halpern's *Flashes of Creation* introduces us to the human side of the Big Bang: the debate about whether it happened and the effort to develop a consistent mathematical description of the early universe. Flashes of Creation is an engaging look at an important moment in the history of cosmology, and in the era of big data and large, diverse collaborations, it is a valuable retrospective of a distinctly twentieth-century approach to fundamental physics."

—CHANDA PRESCOD-WEINSTEIN, author of *The Disordered Cosmos*

"An astute and deeply researched account of the argument between two of the most colorful characters in twentieth-century science about the true nature of literally everything in existence. *Flashes of Creation* shows scientists at their most human, as they struggle to unravel riddles of cosmic importance."

—PHILIP BALL, author of *Beyond Weird: Why Everything You Thought You Knew About Quantum Physics Is Different*

"Beautifully written and thoroughly researched, *Flashes of Creation* offers a fresh look at the mid-twentieth-century debate—sometimes playful, mostly contentious—on the origin of our universe. Halpern's engaging narrative and rich portrayals of the key participants adds a new and reverberating bang to the story of the Big Bang theory's triumph."

—MARCIA BARTUSIAK, author of *The Day We Found the Universe* and *Black Hole*

"A vivid and gripping tale of the origins of today's cosmology."

—IAN STEWART, author of *Calculating the Cosmos*

"Paul Halpern adroitly weaves together the careers of two maverick scientists in this illuminating history."

—FRANK CLOSE, University of Oxford and author of *The Infinity Puzzle*

"Highly readable, entertaining, and informative, Paul Halpern's latest book shows that even when grappling with some of the biggest questions, science is a human activity and all the better for it."

—JAMES KAKALIOS, University of Minnesota and author of *The Physics of Everyday Things*

FLASHES OF CREATION

GEORGE GAMOW, FRED HOYLE, AND THE
GREAT BIG BANG DEBATE

PAUL HALPERN, PHD

BASIC BOOKS
New York

Basic Books
Hachette Book Group
1290 Avenue of the Americas, New York, NY 10104
www.basicbooks.com

Printed in the United States of America
First Edition: August 2021

Published by Basic Books, an imprint of Perseus Books, LLC, a subsidiary of
Hachette Book Group, Inc. The Basic Books name and logo is a trademark of
the Hachette Book Group.

The Hachette Speakers Bureau provides a wide range of authors for speaking
events. To find out more, go to www.hachettespeakersbureau.com or call
(866) 376-6591.

The publisher is not responsible for websites (or their content) that are not
owned by the publisher.

Print book interior design by Linda Mark.

Library of Congress Cataloging-in-Publication Data
Names: Halpern, Paul, 1961– author.
Title: Flashes of creation : George Gamow, Fred Hoyle, and the great Big bang
  debate / Paul Halpern.
Description: First edition. | New York, NY : Basic Books, 2021. | Includes
  bibliographical references and index.
Identifiers: LCCN 2020054994 | ISBN 9781541673595 (hardcover) |
  ISBN 9781541673618 (ebook)
Subjects: LCSH: Gamow, George, 1904–1968. | Hoyle, Fred, 1915-2001. |
  Big bang theory. | Cosmology—History. | Scientists—Biography.
Classification: LCC QB991.B54 H35 2021 | DDC 520.92 [B]—dc23
LC record available at https://lccn.loc.gov/2020054994

ISBNs: 978-1-5416-7359-5 (hardcover), 978-1-5416-7361-8 (ebook)

LSC-C

Printing 1, 2021

Dedicated, with love, to Aden, Eli, and Felicia

Said Hoyle, "You quote

Lemaître, I note,

And Gamow, well, forget them!

That errant gang

And their Big Bang—

Why aid them and abet them?

"You see, my friend,

It has no end

And there was no beginning.

As Bondi, Gold,

And I will hold

Until our hair is thinning!"

—BARBARA GAMOW, "Commentary on Ryle Versus Hoyle,"
from *Mr. Tompkins in Paperback*

Contents

Introduction

THE QUEST FOR THE ORIGIN OF EVERYTHING

> I, for 'twere absurd
>
> To think that Nature in the Earth bred Gold
>
> Perfect i' the instant. Something went before.
>
> There must be remote Matter.
>
> —Ben Jonson, *The Alchemist*

RESOLVING THE GREAT COSMOLOGICAL DEBATE OF THE MID-twentieth century was not on their agenda. Yet in 1964, astrophysicists Arno A. Penzias and Robert W. "Bob" Wilson unexpectedly discovered a radio hiss that turned out to be relic radiation from the early universe. Much to their surprise, their finding, after being interpreted and published the following year, helped settle a long-standing argument about time and space. The Big Bang theory postulated the universe had been created with an initial burst of matter and energy, whereas the steady-state theory—its main rival—described no primordial eruption but rather a slow, continuous creation of material that remains ongoing. The Penzias-Wilson discovery of background radiation tipped the scale toward the Big Bang, away from the steady-state.

Though many researchers had contributed to the development of each theory, in the public mind the debate came down to a clash between two extraordinarily brilliant—and charmingly quirky—figures. Since the late 1940s, Russian Ukrainian American physicist George Gamow—a master of exceptional insights and outrageous puns—had carried the banner of the Big Bang (though he didn't like that expression), and British astrophysicist Fred Hoyle—known for his stubborn persistence, maverick ideas, and passion for long-distance hiking—had tenaciously advocated the steady-state alternative. Their cogent arguments, featured in popular media such as *Scientific American* and the *New York Times*, stimulated ample discussion among savvy followers of science about the possibility or impossibility of a single creation moment in the distant past.

Questions about the origins of everything have a venerable history. Has the universe existed forever? Or was there a beginning? Was all matter and energy created slowly over time, in tiny trickles, or all at once, in a single burst? Are the galaxies in space an evenly distributed mixture of young and old, or is their arrangement a kind of timeline for when they were formed and how they developed?

Long before cosmologists took such questions seriously, they were the province of theologians and philosophers. Pick (or be born into) your religion, and that determined your favored cosmogony. Many ancient belief systems, such as Hinduism, Taoism, and the faiths of the Babylonians, the Greeks (in the time of Plato), and most traditional Native American groups, embraced the notion of cosmic cycles. In the life of the universe, nothing ever truly died. The death of one epoch invariably was followed by the birth of a new one.

The Abrahamic faiths—Judaism, Christianity, and Islam—on the other hand, advocated a single universal creation at some point in the past. The creation moment represented the dawn of humankind and all mortal things, in stark contrast to the concept of the eternity of God. Like the lives of those fated to grow old, get ill, and die, such a unidirectional, linear time scheme started with a clear, glorious birth.

By the 1920s, and especially in the decades following, thanks to the work of Albert Einstein, Georges Lemaître, Edwin Hubble, and other scientists, the debate as to whether or not the universe had a beginning had shifted to secular grounds. Einstein's theory of general relativity precisely laid out how mathematical models could trace the development of the universe. Lemaître and others (notably Russian mathematician Alexander Friedmann) used Einstein's system to provide scientific timelines of cosmic events. Lemaître himself speculated that at a finite time in the past the universe began in an extremely compact, ultra-dense state—something like a giant atom—and grew until it reached its present size. And Hubble, drawing heavily on the work of astronomers Vesto Slipher, Henrietta Leavitt, and others, used the Hooker telescope at Mount Wilson Observatory in California to demonstrate that the universe is full of galaxies and that all but the closest are receding (moving away) from our galaxy, the Milky Way, at a rate that depends on their distance. Lemaître argued that Hubble's result supported his prediction of an expanding universe, and Einstein eventually came to agree. Hubble remained agnostic about whether or not the universe was expanding, simply pointing to his galactic recession data as enough of a contribution to the discourse.

Given Hubble's ambivalence, Einstein's interest in other topics, and Lemaître's humble reluctance to trumpet his own ideas (he was a priest as well as a scientist), scientific cosmology was little publicized in the years leading up to and including World War II. There was no real public debate, simply reports in the pages of scholarly journals.

That all changed in the late 1940s, thanks to Gamow and Hoyle. Each, in his own way, was a polymath, a rebel, and a master at science communication. Each was a great fan of Hollywood and its magical theatrics, recognizing the potential of new media (radio and television) to convey extraordinary ideas to wide audiences. It gave them power to sway the public toward their ideas, lending them an impact well beyond the reach of scholarly journals and specialized popular magazines.

Russian Ukrainian American physicist George Gamow, who argued that the early universe was a hot, dense cauldron of chemical elements in formation. CREDIT: AIP Emilio Segrè Visual Archives, Physics Today Collection.

Gamow, a nuclear physicist who came to embrace cosmology, took up the mantle of Lemaître's idea. Along with his student Ralph Alpher and another researcher, Robert "Bob" Herman, he wrote several key papers on the topic of cosmogenesis—the creation of matter in the fiery early moments of the universe. Gamow's offbeat humor and keen sense of how to elucidate the strange aspects of modern science enlivened his work and made it very palatable to the general public. No solemn clergyman was he; rather, he loved to clown around like a late-night television host. Pranks and practical jokes were his trademark. In numerous popular works, such as his delightful series of books about a bank clerk named Mr. Tompkins who encounters various scientific wonders, his ample wit and zany illustrations made science tons of fun. Gamow's vivid description of cosmic expansion—from an ultradense point to its current size—in the 1952 book *The Creation of the Universe* truly brought what came to be known as the Big Bang theory to life. In that and other works, he used a rather silly-sounding word, "ylem," to

British astrophysicist Fred Hoyle, co-developer of the steady-state theory of cosmology and proposer of the notion that most of the chemical elements arise in the hot cores of stars. CREDIT: Photo by Ramsey and Muspratt, courtesy AIP Emilio Segrè Visual Archives, Physics Today Collection.

describe the primordial ultradense state of the universe. *Ylem* (a term chosen by Alpher) comes from the obscure medieval Latin term *hylem*, meaning "matter."

Fred Hoyle, the leading advocate of the steady-state theory, wasn't quite a showman in the same way as Gamow. His sense of humor was drier, and much less obvious. While Gamow would grab for the silly joke or outlandish pun, Hoyle exercised a darker, more cynical wit. For example, he would jest about cruel circumstances in life, such as the challenges of aging.[1]

Hoyle often used ridicule to disarm his intellectual opponents. He would frequently frame other researchers' theories in a less-than-flattering manner, in a way that made his own ideas sound much more sensible. In those cerebral sword fights, he virtually never derided the opposing scientists themselves, many of whom, such as Lemaître, he deeply admired. (The major exception was his Cambridge colleague Martin Ryle, whom he disliked for his volatile personality and arrogant, disrespectful attitude, as well as for his rival ideas.) Rather, in the spirit of a college debate team champion, he would respect his opponents while finding weaknesses in their arguments. Hoyle's clever critiques, like Gamow's buffoonery, made for great press.

With his dagger-sharp wit, Hoyle deemed patently ridiculous the idea that all of the material in space was created at once sometime in the past. He saw it as a cheap trick—a sleight of hand unbecoming of serious scientists. He coined the term "Big Bang" as a mocking epithet during a BBC science radio program first broadcast on March 28, 1949: "These theories were based on the hypothesis that all the matter in the universe was created in one big bang at a particular time in the remote past," he said. "It now turns out that in some respect or another all such theories are in conflict with the observational requirements. . . . Investigators of this problem are like a party of mountaineers attempting an unclimbed peak."[2]

Hoyle's final remark, about mountain climbing, was even harsher than it sounded because he prided himself as an expert hiker and climber. By the end of his life, he was delighted to have "bagged" all the Munros, the highest mountains in Scotland: 282 peaks all over three thousand feet.[3] So, in effect, he was saying, "Leave cosmology to the experts, or study harder and become a real one yourself."

Hoyle maintained a steady presence on British radio as the leading critic of the "Big Bang," including in a five-part series, "The Nature of the Universe," that was published in book form in 1950. He also became a prolific writer of popular science and science fiction. Like Gamow, he became known for his popularization of science in addition to his scientific accomplishments themselves.

Gamow never liked the term *Big Bang*, because he considered the birth of the universe to be neither big (the universe was tiny at that point) nor a bang (the universe never really exploded; space just grew). Nevertheless, the name stuck.

Likely because of the strength of their personalities and their widespread media presence, Gamow and Hoyle were seen as the main opponents in the epic debate between the Big Bang and steady-state theories. Never mind all the other major contributors, from Friedmann and Lemaître to Alpher and Herman, on the Big Bang side, and Hermann Bondi and Thomas Gold, co-proposers along with Hoyle, on the steady-state side. Although highly informed readers knew of all the others, those following the story in the popular press, particularly in the United States, focused mainly on the two most prominent. During an oral history interview conducted by Martin Harwit, Alpher agreed that the contest "was usually put in the context of Hoyle versus Gamow."[4]

From the early 1950s to the mid-1960s, enthusiasts of popular science had fun taking sides in the great cosmological debate. Whereas Gamow and Hoyle were resolutely secular, focusing only on the science, some members of the public tied the ideas to matters of faith. As in many scientific disputes, until all of the facts come in, personal, philosophical, and religious preferences often prevail. That's what made the Big Bang versus steady-state battle so compelling. In the former, time and the universe have a definite beginning. In the latter, time and the universe are eternal. Consequently, many religious people saw in the Big Bang evidence of divine creation. People who would rather have no need for a creator generally gravitated toward steady-state. It was a choice based on belief rather than evidence—a parlor wager that made sense only before the Penzias-Wilson discovery of the cosmic radio hiss would tip the scale.

Starting in 1964 and continuing into 1965, using a device called the horn antenna at Bell Laboratories in Holmdel, New Jersey, Penzias and Wilson found persistent radio noise that appeared the same coming from all directions. No matter which way they aimed the detector, it was there. After ruling out all manner of local causes, they were greatly puzzled. Fortunately, a team from nearby Princeton University, headed by Robert H. "Bob" Dicke and including young theorist P. J. E. "Jim" Peebles, deduced the answer: the hiss was relic radiation from the hot fireball of the early universe, cooled down to a frigid temperature of approximately 3 degrees above absolute zero. Their results changed the course of science, rendering—at least in the minds of most mainstream researchers—the Big Bang theory of the universe established fact and the steady-state theory a historical curiosity. (Alpher and Herman, Gamow's associates in proposing the Big Bang, had predicted such leftover radiation.) For their extraordinary contributions, Penzias, Wilson, and (most recently) Peebles would each win a Nobel Prize.

But neither Gamow nor Hoyle were one-trick ponies. Their cosmological disagreement was only one facet of their extraordinary contributions to science, as well as their popularization of it. Reveling in literature and the arts and such diverse scientific fields as genetics and astrobiology, they were arguably two of the most creative scientists of the twentieth century. Despite very different backgrounds, they were each raised in a similar fashion: to take joy in the process of discovery rather than reveling only in the results.

Gamow's and Hoyle's brilliance cast light on another age-old puzzle: how the diversity of elements in the cosmos emerged from more rudimentary components. Thanks to their independent efforts—which turned out to complement each other beautifully—we now know how each element on the periodic table came to be, from simple hydrogen to the more complex higher elements.

The tiny nucleus of an atom is made of positively charged protons and neutral neutrons surrounded by a cloud of negatively charged electrons. The composition of atomic nuclei is what distinguishes one element from another. The most basic type of nucleus—that of the most common kind of hydrogen atom—includes a single proton. The nuclei of other elements are plump with different numbers of protons and neutrons. For example, the most common variety of uranium, the heaviest naturally occurring element, has 92 protons and 146 neutrons.

Parsimony suggests that complexity stems from simplicity. Rather than purporting that each element—composed of the same protons, neutrons, and electrons as all others—had a wholly independent origin, it makes more sense to suppose that some natural process of combining nuclei built up light elements, such as hydrogen, into heavier ones, such as helium, lithium, and so forth, finally leading to formation of the bulkiest ones, such as uranium. Yet, as it turned out, designing a viable model that explained how all the natural elements emerged from simpler ones was not easy.

Research in the 1920s and 1930s by brilliant scientists such as Arthur Eddington and Hans Bethe (aided by a key insight of Gamow) showed how two hydrogen nuclei could combine to form helium; this fusion process powered the sun. But explaining how the lion's share of chemical elements in space had emerged proved daunting.

Fortunately, two extraordinary minds were up to the challenge. Thanks to the genius of both Gamow and Hoyle, we finally know how the atomic nuclei of all of the elements in the periodic table are forged. Independently, these scientists demonstrated how, in two distinct ways, nature's construction processes turn simple building blocks into complex structures via nuclear fusion taking place at temperatures much hotter than those at the core of the sun.

Gamow, along with Alpher and Herman, developed a scheme called Big Bang nucleosynthesis in which the nuclei of the known chemical elements were built up step-by-step from simpler nuclei during the first few minutes of the ultradense early universe. Because of the universe's initial concentration of energy, the temperature in the nascent moments of creation must have been extremely hot, which they supposed would enable all of the chemical elements to be forged.

As a nonbeliever in the Big Bang, Hoyle was driven to find another means by which the chemical elements had been formed. He developed and refined the idea—later with William Fowler, Margaret Burbidge, and Geoffrey Burbidge—that the chemical elements are created in the cores of stars in several distinct processes that transpire at different stages of stellar life, including during the sudden contraction of the core as a star reaches its endgame. The newly created elements are then released into space during a supernova burst, with the heaviest of them created during the intense heat of the blast itself. Once dissipated, the heavier elements are available as ingredients for inclusion in new stars and planets, which is why Earth is rich with elements such as nitrogen, oxygen, carbon, iron, nickel, and so forth—not just the lightest elements hydrogen and helium.

Remarkably, both Gamow's and Hoyle's teams were partially right. Most of the helium in the universe, as Gamow, Alpher, and Herman anticipated, was produced during the Big Bang; stellar nucleosynthesis, Hoyle's theory, can't explain the large quantities that exist. On the other hand, Gamow's group never could explain how the higher elements could be produced in the Big Bang. Their main problem in constructing such a stepladder of development was the instability of a key rung: beryllium-8, an isotope with four protons and four neutrons that has a lifetime of only three-hundred-billionths of a nanosecond before it decays into two helium-4 nuclei. And without that critical foothold, they couldn't climb higher on the ladder to reach carbon and even heavier elements.

It was Hoyle's brilliance and persistence that helped find a way around the missing rung—not through the Big Bang but through the immensely hot shrinking cores of dying stars. He was led to that conclusion by his conviction that astrophysical processes must offer an explanation for how such a vital element as carbon formed. Stanford astrophysicist Robert V. Wagoner, who collaborated with Hoyle, noted: "Hoyle broadened our view of possibly relevant physical processes that could help us understand various aspects of our universe. He motivated many theorists to 'think outside the box.'"[5]

Hoyle became aware of the "triple alpha process," suggested by Edwin Salpeter in 1952, that allowed for beryllium-8 to merge with a helium-4 nucleus in the fraction of a nanosecond before it disintegrated. The combination resulted in carbon-12. Salpeter had speculated that this could happen at temperatures above 100 million degrees Kelvin through chance collisions of the atoms. However, he didn't detail how the production of stable carbon-12 nuclei could reliably occur, beyond chance collisions.

That's where Hoyle's exceptional insight came into play. Nuclear physics, governed by quantum rules, makes certain types of transitions more likely, and others less likely or even impossible. Hoyle, working backward from the heavier element, predicted that carbon-12 must have a hitherto unknown energy level equal to that of the combined energies of beryllium-8 and helium-4, thus making the transition of the two into the heavier one much simpler.

It is like constructing a bridge high above the ground between two skyscrapers: it's simplest to do if each building has a certain level of exactly the same height. Thus, if you see a pedestrian bridge connecting buildings you've never been in, you might suspect that they share floors with that height, at least approximately. Alternatively, even if you haven't seen a bridge but rather observe someone walk into the lobby of the first building and, sometime later, exit the lobby of the second, you might suspect that the buildings possess a common level that allows for a connection. Similarly, Hoyle inferred that carbon-12 must have an energy level, undetected up to that point, that allows for ready transitions from beryllium-8 amalgamated with helium-4. Those isotopes would need such a "bridge," he thought, to explain their metamorphosis from one state into another.

The furnace that would allow for such transformations, Hoyle further surmised, was the shrunken core of a swollen red giant star. Once a massive star exhausts its hydrogen fuel, its core cannot produce enough radiation to counteract the gravitational pressure of its bulk, and it starts to collapse. The shock waves of the collapse induce the star's outer envelope to expand into a much larger star—a red giant. The core's collapse causes it to heat up above the requisite temperature of 100 million degrees that Salpeter had suggested, supplying the conditions for the higher elements such as carbon-12 to form from the fusion of lighter elements.

Hoyle spent considerable time at the Kellogg Radiation Laboratory associated with Caltech, first in collaboration with Fowler, and later also with the two Burbidges, to search for such a carbon-12 resonance (conditions under which the energy level matches that required for the formation of carbon-12) and to develop a detailed model of how other higher elements could be built up in the core of a star or forged in the heat of supernova explosions and then released via such bursts. The four researchers published their detailed model in an instrumental paper, "Synthesis of the Elements in Stars," in 1957.

As astrophysicist Virginia Trimble notes, "The paper was so influential that generations of astrophysicists called it B²FH for short, quipping that the early Universe made hydrogen and helium, but Burbidge, Burbidge, Fowler and Hoyle made all the rest."[6]

Because of their contributions to the two different schemes explaining element creation—Big Bang nucleosynthesis for the lighter elements and stellar nucleosynthesis for all of the others—all of those involved, including Gamow and Hoyle but also Alpher, Herman, the Burbidges, and Fowler, were arguably qualified to receive Nobel Prizes. The fact that the only one who did was Fowler has been a controversial matter ever since. One complicating factor is that the Nobel Prize in Physics can be awarded to only three individuals at a time. Another complication is that some of the predictions of Gamow, Alpher, and Herman were initially overlooked at the time of the discovery of the cosmic microwave background radiation (CMBR) that served to verify the Big Bang, with the focus on the calculations by Princeton physicists instead. Thus, it took some time for proper credit to be allocated to everyone involved, making the history rather nuanced. We'll explore some of the conflicts and controversies that arose in the pages ahead.

———

THIS BOOK IS A HIGHLY UNUSUAL SCIENTIFIC JOINT BIOGRAPHY, BEcause the principal players, Gamow and Hoyle, did not interact very often in person. Their most notable face-to-face encounter took place in the summer of 1956 when Gamow was working as a consultant for the defense company General Dynamics in La Jolla, California, and he invited Hoyle for a visit. While driving along the seaside enclave's sunny streets in Gamow's Cadillac, they had an animated discussion about the temperature of space. Their discourse anticipated, in some ways,

Penzias and Wilson's discovery of the cosmic microwave background radiation. In general, though, the two scientists largely ran in different circles. Sometimes the separation was deliberate. In at least one instance when an in-person, scholarly debate could have happened, the 1958 Solvay Conference in Belgium, Gamow sensed he was excluded because of his opposition to Hoyle's ideas. And, sadly, because of his poor health, Gamow died much earlier than did Hoyle.

Yet their lives were tied together, for a time, in so many ways. It was a modern media conjunction of personalities, appropriate because they both loved Hollywood movies, speculative literature, and drama. They met on the pages of numerous accounts of the battle over the properties of the universe, including a special issue of *Scientific American* called "The Universe," published in September 1956, in which their articles, "The Evolutionary Universe," by Gamow, and "The Steady-State Universe," by Hoyle, appeared in succession. Their dueling models of element formation turned out to complement each other like yin and yang. Moreover, their popular books and articles vied for the minds of all those eager to learn science in an accessible way.

A humorous anecdote from around that period shows how the two were closely associated in the public mind. Once, during a scientific conference, Gamow was sitting in a hotel bar having some drinks. Deciding to play a joke on him, another scientist bribed a waitress to approach his table and say, "There is a telephone call for you Professor Hoyle." Without batting an eye, Gamow replied, "Don't throw Hoyle on troubled waters."[7]

In 1959, around the height of the Big Bang–steady-state debate, scientist and critic C. P. Snow, in his highly influential Rede Lecture (an annual series put on at Cambridge—not too far from where Hoyle was working at the time) spoke of the great divide between "two cultures": the scientific realm and the literary world.

"I believe the intellectual life of the whole of western society is increasingly being split into two polar groups . . . at one pole we have the literary intellectuals . . . at the other scientists," Snow argued. "Closing the gap between our cultures is a necessity in the most abstract intellectual sense, as well as in the most practical. When those two senses have grown apart, then no society is going to be able to think with wisdom."[8]

Hoyle, who would deliver a Rede Lecture on cosmology twenty-three years later, demonstrated that not all intellectuals fell into only one of Snow's two cultures. Throughout his life, he argued strongly that scientists should be literate, proving his own thesis by writing or cowriting numerous well-regarded science fiction books that blended thought-provoking scientific ideas with intriguing social issues. For example, his novel *The Black Cloud* and television screenplay (and novelization) *A for Andromeda* offered two distinctive fictional reflections of what alien life might be like. Moreover, he often ventured into the world of the arts—writing the libretti of an opera, for example, and an oratario with the composer Leo Smit.

Gamow was not inclined to make political or social statements, such as calling for scientists to be more literate. Nevertheless, in his deeds, he similarly set a great example of a physicist savvy about culture. His numerous popular books and articles—Snow, in his later role as editor of *Discovery* magazine, would help launch him as a writer—contained clever sketches and word play. In a book about the history of quantum physics, he included an English translation of a parody of Faust performed in Copenhagen to gently poke fun at the senior scientists at Niels Bohr's Institute for Theoretical Physics. He illustrated it himself with hilarious caricatures. Even in his scientific articles, Gamow made erudite jokes, such as humorously claiming Bethe "in absentia" as the third author of one of his key papers with Alpher, just so that the byline could read "Alpher, Bethe, and Gamow," which sounds like "alpha, beta, and gamma," the first three letters of the Greek alphabet.

If, during the UFO craze of the 1950s, an extraterrestrial race wanted to search the earth for some of its finest Renaissance men, individuals adept at science and in the arts, with brilliant instincts about how the universe works, it could do no better than select Gamow and Hoyle.

These two thinkers seemed otherworldly indeed, each with an extra-ordinary intellect and buoyant imagination that verily rocketed them to the stars and beyond.

Following one's gut instincts has a downside, however. Impulsive behavior arguably served to marginalize both Gamow and Hoyle in their later years, each in a different way. Gamow often introduced ideas but didn't follow up on them. Rather, he'd leave the work to others and simply move on. His unabashed, joking manner sometimes overwhelmed his colleagues, who might have taken him less seriously during research discussions than they should have. His low self-control manifested in chain smoking and heavy drinking, taking a toll on his health and (especially the latter habit) affecting how others perceived him. Those who valued his enormous contributions worried he'd be seen as a drunken clown rather than as the genius he was.

Hoyle, on the other hand, lived a healthy lifestyle but sometimes made poor choices in other ways. In his final decades, he spent much of his time on projects that were well out on the fringes of conventional science, claiming, without evidence, that life on Earth was brought here by astral bodies such as comets and alleging that a famous fossil in London's Natural History Museum was fabricated. Rattled by politics at the University of Cambridge, he unexpectedly resigned from his academic position only a few years before he was due to retire and moved with his wife to the Lake District, a remote part of England, which served to isolate him from other scientists. Finally, his repeated dismissal of all evidence that the observable universe was once hot and compact, while contriving far-fetched alternative explanations, raised many eyebrows and made it hard for the mainstream scientific community to continue to take him seriously, despite his pivotal earlier contributions.

Arguably in both cases, each man's reputation as an outstanding popularizer with a fanciful side also made him suspect in the eyes of serious scientists. As Snow pointed out, some scientists were uninterested in bridging the divide between the "two cultures" and belittled those who made such attempts. Gamow's popular accounts could be very silly, and Hoyle's science fiction could be rather outlandish—rendering their works fun for readers but odd for certain hard-headed researchers. Passionate about creativity, neither Gamow nor Hoyle cared what traditionalists thought. They spoke to a wider audience and to a higher principle: the search for and dissemination of truth.

Fundamentally, Gamow and Hoyle were adventurous loners who cared far more about cosmic mysteries than social conventions. Each detested bureaucracy, which they saw as holding back individual creativity. From childhood until the end, each followed his own path through the terrain of scientific discovery, even during times when he was outcast from the scientific community. Solitary and stubborn, each found truth and joy on his own terms and never wanted to be part of the herd.

CHAPTER ONE

Children of an Expanding Cosmos

In one corner you have burly, pun-making Russian-American physicist George Gamow. He says the universe did have a beginning and that beginning was a very big bang....

In the other corner you have piano-playing, novel-writing, baggy-tweeded English astronomer Fred Hoyle. His side says that there was no instant creation. The universe is in a steady-state.

—Martin Mann, "The March of Science," *Popular Science*, March 1962

A CHILD VYING WITH THE UNIVERSE IS AN UNEVEN MATCH FOR sure. Fledgling youth are small and helpless. The sky seems limitless and overwhelming. Yet stubborn, brilliant children—those with just the right mixture of audacity and insight—cede no ground. Schoolyards and playgrounds are too confining. Even Earth, with its ancient habits and tired superstitions, seems trite to those with wide eyes and novel perspectives. Seize the cosmos with the mind, explain its glory through reasoning and imagination, and the intrepid soul wins the ultimate prize.

In the early twentieth century, science fiction and science popularization—through the works of visionary writers such as Jules Verne,

Camille Flammarion, and H. G. Wells; pioneering film directors such as Georges Méliès (A *Voyage to the Moon*) and George Booth (*The Airship Destroyer*); and the first radio broadcasters, which aired vivid, groundbreaking theatrical productions—sparked youthful imaginations in an unprecedented manner. The era of emerging mass media, which publicized scientific discoveries of the age like never before, would prove the perfect nurturing ground for two extraordinary individuals, each with an unstoppable drive to tackle the great questions of science and convey their bold hunches and radical notions about the universe to wide, eager audiences.

The epoch of scientists popularizing their own work—for good or bad—had commenced. No longer would theories and hypotheses remain hidden within the pages of scholarly books and journals until clear-cut experimental findings rendered them valid and relevant. At its best, the publicity of science led to greater public understanding. But at its worst, media-amplified science allowed people to mistake wild speculation for established fact—as happened in 1910, when apocalyptic fear arose around the approach of Halley's Comet, sparked in part by Flammarion's well-publicized conjecture that a poisonous gas carried by the comet threatened life on earth. For better or worse, our two protagonists grew up in, and became attuned to, this media-driven age.

IN THE DAYS OF THE COMET

George (Georgiy Antonovich) Gamow was born in Odessa in the Russian Empire (now Ukraine) on March 4, 1904. He came into the world rather dramatically. As a fetus, he was huge and misaligned in the womb. Thus, when his mother, Alexandra Arsenievia Lebedinzeva, a high school teacher, went into labor in the family's apartment late at night, her life was gravely in danger. Luckily, their next-door neighbor knew of a good Moscow surgeon who happened to be vacationing nearby. In her horse and buggy, the neighbor sped off into the darkness to the doctor's house, roused him from his sleep, and fetched him to

German-born physicist Albert
Einstein, developer of the
special and general theories of
relativity. CREDIT: AIP Emilio
Segrè Visual Archives.

perform an emergency Caesarian section. The procedure took place on
a desk next to bookshelves—which Gamow came to think of as a fitting
place for a future writer to begin his life.[1] Hefty baby George—who
would grow up into a towering man—was fine, thankfully.

George's father, Anton Mikhailovich Gamow, was also a teacher,
specializing in Russian language and literature. Perhaps Anton Gamow's
most famous pupil was Lev Bronstein, better known by his nom de
guerre Leon Trotsky, the Russian revolutionary who worked closely
with Lenin and who would eventually become Stalin's chief opponent.
A disgruntled student, he circulated a petition to try to get his teacher
fired for alleged incompetency. Luckily, Anton survived the attempted
"Trotskyist coup," remained an educator, and continued to fund his
family's ample book collection and other necessities of life.

The year after Gamow was born was a banner year for Albert Einstein, whose theories would play a major role in Gamow's life. In 1905,
Einstein published several revolutionary papers, including ones that laid

out his special theory of relativity. In a flash, he shattered the rigid Newtonian framework of absolute space and time that had been canon for
two centuries. Absolute space imagines a set of invisible rulers that spans
a fixed universe in all three dimensions. Absolute time does the same for
the passage of moments, imagining them clocked linearly in a consistent
way throughout the cosmos, steadily ticking onward through eternity.

Einstein greatly respected Isaac Newton, so he would not have banished absolute space and time for trivial reasons. Rather, he wanted to
resolve a stark contradiction in the science of light between the predictions of classical mechanics, as described by Newton and others, and
those of classical electrodynamics, as delineated by the equations of the
Scottish physicist James Clerk Maxwell.

As a teenager, Einstein constructed a thought experiment about an
imaginary person racing alongside a beam of light. Classical mechanics
predicts that such an incredibly swift runner would perceive the light
beam as frozen in space as they keep pace with each other. After all,
that's what happens when two trains move at the same steady pace in
the same direction on parallel tracks. Looking out their windows, passengers on each train view those sitting in the other train as being at rest.
So why not the same observation if we could keep pace with a beam of
light traveling through space? The problem was, as Maxwell's equations
demonstrated, light's speed in a vacuum does not depend on the speed
of the observer, as Newtonian physics predicts. Rather, it always remains
the same no matter how fast the observer is moving. In contrast to the
train situation, even if someone tried to race a light beam, he or she
couldn't possibly catch up and view it as being at rest. Frustratingly,
the light would always seem to be dashing through space at exactly the
same speed.

Resolving the contradiction between Newtonian and Maxwellian
physics, Einstein found, required making measurements of intervals in
time and space dependent on the relative speeds of the person taking
those readings and the phenomena she or he is observing. Such flexibility in describing time and space, he discovered, would allow for a

self-consistent description of motion that encompasses a constant speed of light. That is, by relaxing the rigidity of clocks and yardsticks in Newtonian physics, Einstein was able to rescue its general picture of relative motion for everything except light, while supporting the Maxwellian notion that light itself has a fixed velocity.

In special relativity, observers traveling at different speeds record time passing at different rates—an effect increasingly noticeable if their relative velocities approach the speed of light. For example, when a bystander on Earth observes (via an extraordinarily powerful telescope) a timepiece on a spaceship traveling close to light speed, the seconds seem to tick by more slowly than they do on a stationary clock hanging on the wall. That effect is called time dilation. Moreover, the moving vehicle appears to shrink in the direction of its motion, an effect called length contraction. Speed equals distance divided by time. Therefore, by showing that distances and times are observer-dependent, Einstein made it possible for all observers to measure the same speed of light in a vacuum, while applying laws of relative motion that govern all objects. Light speed (absolute) and the velocities of trains (relative to the speed of observers), for example, could thus be encompassed in a self-consistent way. In essence, special relativity brilliantly resolved the discrepancy between the constancy of light and the relativity of motion slower than light by dismissing the idea of universal rulers and clocks and introducing flexible definitions that depend on the relative speed of the observer and what is being observed.

Such malleability of space and time has been confirmed in numerous experiments involving high-speed objects, such as in precise measurements of the decay rates of elementary particles propelled by accelerators close to the speed of light. Even in the case of much slower speeds, such as passengers traveling in aircraft, scientists have been able to measure a minuscule, but well-documented, effect. Their clocks have ticked at a slightly slower pace than those on Earth, rendering them tiny fractions of a second younger than if they had remained on the ground.

In a flash, Einstein had banished the Newtonian absolutist frame-work and thereby made physics more internally consistent. His 1905 works, including those introducing special relativity, were so extraordi-nary, historians have deemed the year his *annus mirabilis*, or "miracle year."

Of course, at the age of one, Gamow had no idea what was happen-ing in the world of science. Yet, he would soon catch up. As he passed through childhood, immersed in his parents' impressive collection of books, he became a true bibliophile. At first his mother would read to him, then he became a voracious reader himself. Influenced by writers such as Jules Verne, whose novel *From the Earth to the Moon* galva-nized Gamow's imagination, and Flammarion, a popular astronomer who produced mind-stretching, marvelously illustrated works, Gamow soon turned his fascination to astronomy. Such speculative writings planted the seeds for his interests in science much later in life.

One memorable event in Gamow's early life, and indeed for the entire world, was the close passage of Halley's Comet. Earth crossed quietly through its tail on May 19, 1910. Despite the innocuous en-counter, its approach had fomented a great deal of fear. A report had surfaced that researchers tracking the comet from the Yerkes Observa-tory in Wisconsin, which was operated by the University of Chicago, had found spectral evidence of the chemical cyanogen in its makeup. In large doses, cyanogen is a deadly poison. Flammarion, who had en-visioned a comet-caused apocalypse in his 1894 book *Omega: La Fin du Monde* (*The Last Days of the World*), warned that "the cyanogen gas would impregnate the atmosphere and possibly snuff out all life on the planet."[2]

Odessa was not immune to the ensuing panic. Aside from the wor-ries about "poison gas" (which was in truth so dilute as to be absolutely harmless), unusual astral phenomena often generate fear. In folk belief, comets are harbingers of catastrophe. In Odessa, and the southern Rus-sian Empire in general, unscrupulous religious individuals took advan-tage of public worry to hawk special protection prayers for money.[3]

Yet when the comet arrived, it turned out to be a thing of beauty and wonder, not terror. Gamow climbed on the roof of his family's house to gaze at its splendor. He decided soon thereafter (inspired more by the books he read than by the comet) that he wanted to be an astronomer, which he later modified to being a physicist. His father bought him a telescope, and young Gamow put it to good use. As an adult, his work bridged both fields.

Sadly, Gamow's mother died when he was nine years old. Raising him alone, his father continued to support the boy's science interests—buying him a microscope, for example—and sought to enrich his musical tastes by taking him often to the opera. Despite his father's prodding, Gamow never developed a taste for that form of musical expression. The one exception was the opera *Russlan and Ludmilla*, based on one of Alexander Pushkin's fairy tales, because the performance featured a particularly memorable scene involving the severed, but still-animated head of a giant, which stirred Gamow's scientific curiosity.

AN AGE OF REVOLUTION

In the early to mid-1910s, the Russian Empire, the German Empire, and the British Empire were ruled by cousins: Nicholas II, tsar of Russia; Wilhelm II, German emperor; and George V, king of the United Kingdom and the British Dominions. (King George, constrained by a parliamentary democracy, played a largely ceremonial role.) Photos taken on family occasions, such as in Berlin at the 1913 wedding of the kaiser's daughter Victoria Louise, highlighted the three sovereigns' striking resemblance. Thus, as subjects of those monarchs, George Gamow, Albert Einstein, and Fred Hoyle—despite living in three widely separate parts of Europe—had more in common than one might think. Revolutionary developments involving their respective empires during World War I and its aftermath would teach these scientists that nothing was constant in the world, scientific theory no exception. They also shared some personality traits—dogged individualism, decision-making

The house in Gilstead, West Yorkshire, England, where Fred
Hoyle spent his early years, now honored with a blue plaque.
CREDIT: Photograph by Paul Halpern.

based on intuitive judgment, distrust of religious dogma, and nonconformity in the face of scientific orthodoxy—that were shaped perhaps by
the turbulence of the age.

Hoyle was born on June 24, 1915, in the village of Gilstead, near
the town of Bingley in the West Riding of Yorkshire (an area that loosely
overlaps what is now called "West Yorkshire"), England. Unlike Odessa,
a seaport, Bingley is well inland. Yet at that time it was open to much
commerce, from sea to sea, thanks to its prominent position on the Leeds
and Liverpool Canal, a major Industrial Age transportation route. One
of the marvels of the canal-building era is Bingley's two sets of locks—
Three-Rise Locks and especially Five-Rise Locks—which helped vessels climb dozens of feet in short intervals of time. In fact, the latter,
with its staircase design, is the steepest in England. In Hoyle's time, as it
does today, it represented the most famous tourist site in Bingley.

Gilstead itself is tiny and cozy, laced with walking paths, meadows,
and rows of working-class houses. The Glen (originally Hammondale),

The Glen, formerly Hammondale, Gilstead, Yorkshire, England, built by acclaimed poet Ben Preston, Fred Hoyle's great-grandfather. CREDIT: Photograph by Paul Halpern.

a pub founded by Hoyle's great-grandfather Ben Preston, a famous poet who wrote in the West Yorkshire dialect, is a central feature. Preston hailed from the nearby city of Bradford, where he worked as a sorter (quality controller) in a woolen mill. There, at the age of thirty, he wrote his moving poem "Come to Thy Gronny, Doy." Once he had saved up enough money, through a successful second career in writing and journalism, he moved his family to the Bingley area, where it remained for generations.

In those days, it was not uncommon for people somehow related to each other to wed. So it happened that two of Ben Preston's descendants—Ben Hoyle and Mabel Pickard—got married. Because their mothers were sisters, Ben and Mabel were first cousins. As they would end up being Fred Hoyle's parents, he had a complex family tree.

Ben Hoyle, like his namesake ancestor, worked in the Bradford wool industry—in his case, as a respected cloth dealer. Mabel, a talented musician, trained at the Royal College of Music and became a school-teacher. When they got married, however, and Mabel took the surname Hoyle, social rules of the time demanded that she give up her career and become a housewife. And she did, for a time.

But less than a year before Fred was born, Europe plunged into World War I, with Britain joining on August 4, 1914. As the war dragged on, more fighters were needed and Ben Hoyle, already in his thirties, was conscripted into the army. He fought for almost three years (from 1915, right after Fred was born, until 1918) as leader of a machine-gun unit. It was a dangerous post, with only a fraction of soldiers surviving more than a few months. Thus, Fred, as a very young boy, was without a father and feared losing him forever.

Meanwhile, Mabel needed to make ends meet. Although women whose husbands were away at war were given special permission to teach, she didn't want to leave Fred by himself the whole day, so she preferred an afternoon or evening job. Even though she was a classically trained pianist, more familiar with Beethoven than ragtime, she applied for a job as the accompanist at a local cinema. As the silent movie reels unraveled, her nimble fingers danced across the keys. When the owner realized that she was playing serious classical pieces rather than the kitschier film music then in vogue, he fired her. Soon he had to bring her back, however, when customers complained. It seems they had largely bought their tickets to hear her performances rather than to watch the films.

Thus, as a wee lad, Fred spent considerable time alone. Did his fiery independence and stubbornness stem from being by himself for so many hours? Or was it in his very nature? An old saying goes, "You can always tell a Yorkshireman—but you can't tell him much."[4] As a lifelong trait, Fred was not easily persuaded to change his views once he had established them. Upon the close of the war, his father returned to Gilstead safely, albeit shell-shocked, and Fred didn't have to become the man of the house at the age of three, as many had feared.

It must have been delightful for the whole family to relax to the strains of Beethoven again. Between listening to his mother perform daily on the piano and later learning to play the violin from his father, Fred acquired a lifelong love of music. Among his many accomplishments, in midlife he collaborated with American composer Leo Smit and wrote the libretti for two operas.

Hoyle's son, Geoffrey, noted, "Music was an integral part of his life from a very early age providing both intellectual stimulus and relaxation. His parents regularly hosted musical soirees at their home."[5]

His taste for science, on the other hand, arose from a natural curiosity. A grandfather clock that his father had repaired shortly after returning from the war stimulated a deep interest in the nature of time. He badgered his parents with the same question, "What's the time?" naturally getting different answers, until finally it dawned on him that their responses were tied to the motion of the hands of the clock.[6] "Science to my father had no boundaries," noted Hoyle's daughter, Elizabeth. "As a young child he learnt things for himself. He approached most subjects with an open mind and most people as well."[7]

Math came easy to him. By the age of four, he was a whiz at multiplication. Reading he picked up several years later by deciphering the title cards of silent films at the cinema. Perhaps because of his mother's occupation, he acquired a passion for movies. He was prone to visual problems, so his love of film might explain why his reading began with large print on a screen.

By 1921, a little sister named Joan arrived. With the two children taking up much of her time, Mabel had her hands full. She fell ill for some time, possibly suffering from postpartum depression.[8]

Later in childhood, Hoyle picked up a love of chemistry by perusing one of his parents' books. His father also had a largely unused chemistry set that young Hoyle found utterly fascinating.[9] He conducted numerous experiments at home, even making gunpowder. Once he could travel on his own, he took a streetcar to Bradford, popped into a wholesale pharmacy, and asked for glass tubing and concentrated sulfuric acid, a highly dangerous chemical that could cause severe burns if mishandled. Yet, he must have seemed, as a boy, to know what he was doing because his request was fulfilled.[10]

Hoyle also became interested in his father's radio receivers.[11] Gilstead maintained a certain isolation from the rest of the world. It was common in that region, a haven for industry, for people to tinker with gadgets. Amateur radio offered a ready, homespun way of connecting with others.

In short, learning for Hoyle was fun. It was formal education that he would come to detest. He much preferred to think for himself, not be told what to do. Autocrats, bullies, and powerful bureaucrats of any stripe revolted him.

After the war, residents were anxious to get back to normal life. Yet revolutions—both political and scientific—boiled up all over the planet. In Russia, Lenin and the Bolsheviks had taken over, precipitating much chaos from Saint Petersburg and Moscow down to Odessa.

At some point, Hoyle came across a reference to (pre-Bolshevik) Russian revolutionaries in a novel, *Under Western Eyes*, by Joseph Conrad. Because of an abusive, traumatic experience he had had in primary school (Mornington Road School), Hoyle identified with one of the main characters, Razumov, who was beaten by thugs, punched on both ears, and rendered deaf.

Hoyle's horrific incident at school started out with a seemingly innocuous assignment. His teacher had asked each pupil to pick a flower and talk about its features. Hoyle plucked one variety and noted that it had six petals. When he brought it into school, his teacher insisted that such flowers had only five petals. Sticking to his guns, Hoyle pointed out that the teacher must not have been counting correctly.

The teacher was furious. He took Hoyle aside and punched him hard on the left side on his head. Memories of the shock and pain reverberated for years. Hoyle began to skip school and take long walks— to Five-Rise Locks, to the meadows and moors—just to get away from that awful place. His mother finally enrolled him in Eldwick School, though it was some distance away, but by then he was an excellent walker. Later in life, Hoyle went deaf in his left ear, quite possibly because of that teacher's blow. He thought of the passage in Conrad's novel: "Razumov . . . received a tremendous blow on the side of his head over his ear. At the same time he heard a faint, dull detonating sound, as if someone had fired a pistol on the other side of the wall. A raging fury awoke in him at this outrage."[12]

Under Western Eyes was set during the time of the failed Russian revolution of 1905. Between then and 1917—especially during the time of World War I—food shortages, a crashing economy, and growing awareness of the gaping inequalities under the czarist regime set the stage for rising protests and ultimately the overthrow of the monarchy. During the turbulent years of the Russian Civil War that followed the Bolshevik Revolution, Odessa represented a prize for both sides of the struggle. The communist Red Army and anticommunists battled it out for years, each vying for dominance over the city.

Those happened to be the years Gamow was in high school, that is, when it was open. Clashes and fuel shortages repeatedly disrupted his schooling. Fortunately, like Hoyle, he was brilliant enough to learn much on his own. For example, whereas his classmates struggled with high school algebra, Gamow found it too simple. Instead, he taught himself differential equations, a topic typically reserved for upper-level university courses.

It was around then that Gamow had one of his first encounters with the British. During a time when food and drinkable water were scarce, vessels from the UK and other countries such as France brought precious supplies to the harbor, which was connected to the main part of the city by the famous long, steep stairway that later (because of the 1925 film *Battleship Potemkin*) became known as the "Potemkin Stairs." The potable water was not pumped up to the city center, however, but could only be accessed via slow public faucets near the docks. Citizens lined up each day for that precious opportunity. One of Gamow's family chores was to bring down buckets, wait in line, slowly fill them with water, and then haul them home up the numerous steps.

One day Gamow was walking near the harbor with pails in hand when he spotted a British sailor standing near his ship. When the seafarer asked him what he doing, he decided to practice his English and responded that he needed some water. No problem, replied the sailor, who quickly filled the buckets using a hose from the ship. Others nearby, who saw how fast and easy it was, also had their pails filled. Gamow lugged the buckets up the long staircase, but it was only at the top that he realized he had been tricked. The water, simply saltwater pumped from the sea, was undrinkable. Rather than curse the British, though, he had a good laugh. It was his first experience with their prankish sense of humor.

As Odessa reeled from the turmoil following the Russian Revolution, world headlines proclaimed a far different kind of revolution: one that overthrew conventional notions of space and time. On May 29, 1919, a total solar eclipse could be seen in parts of the Southern Hemisphere. Two British expeditions, organized by Arthur Eddington and Frank Dyson, had measured the properties of starlight in the mooncovered sun's vicinity to test a prediction of Einstein's theory of gravitation, known as the general theory of relativity. That data was analyzed and reported at a joint meeting of the Royal Society and the Royal Astronomical Society (RAS) held on November 6 of that year. Eddington and Dyson concluded that the results of their studies were consistent with general relativity. In the days that followed, Einstein became world famous. "Revolution in Science" declared the front page of *The Times* of London on November 7, "New Theory of the Universe: Newtonian Ideas Overthrown."[13]

Gamow somehow became aware of Einsteinian relativity during his high school years, and came to embrace it. Precociously, he began to explore its implications. He enrolled at Novorossiya University in Odessa, hoping to study the sciences, especially physics. However, university life there was in shambles because of the Russian Civil War. Much to Gamow's great disappointment, the chair of the physics department decided that the university's lack of personnel and suitable equipment for demonstrations precluded them from conducting reasonable physics lectures, and none were held. The math program was good, but that wasn't enough. After a year, the frustrated Gamow implored his father to let him transfer to a better university. His father sold some of the family silver, which enabled Gamow in 1922, at the age of eighteen, to take the life-changing voyage up to Saint Petersburg (then called Petrograd, and later Leningrad) to attend the excellent university there, with its outstanding physics program.[14]

MOLDING THE COSMIC CLAY

Einstein's journey to fame and accomplishment was years in the making. After introducing special relativity in 1905, he realized the theory was incomplete and began to think about ways to extend it. Special relativity, in its simple, original form, applies to situations in which objects are moving at a constant velocity, not to cases in which an object is accelerating—speeding up, slowing down, or turning. And it doesn't address the effects of gravitation. In short, it is incapable of describing the bulk of phenomena in the natural world.

One issue of gravitation that Einstein believed he needed to address had to do with Newton's concept of gravity as a remotely acting force; instead, Einstein believed it was something more immediate. Newton imagined gravity as a kind of invisible thread of attraction binding Earth to the sun and keeping this planet in orbit. But what if, somehow, the sun blinked out of existence? According to Newtonian physics, Earth would instantly be free to move in a straight line off into space. But sunlight takes about eight minutes to reach Earth. Surely, any kind of effect due to the sun's vanishing must take at least that time to make the leap across the chasm of darkness in between. It would be very strange if Earth starting racing away even before people noticed the sun was missing. In essence, because special relativity precludes faster-than-light interactions, Einstein supposed Newton's model of gravitation as an instantaneous remote effect must be wrong.

Einstein found the way forward by means of a very clever thought experiment involving objects in free fall. He imagined someone falling off the roof of his house and letting go of an object—say, a toolbox—as he begins his descent. As he is plunging, he notices that the object is descending right next to him at the same rate. If he didn't know he and the object were falling, he might believe that they both were at rest with respect to each other. Thus, if you imagine a bubble around the man and the object, within it would be a system that resembles being in the state of inertia. Inertia means an object is either at rest or moving in a straight line with a constant speed. That image led Einstein to the equivalence principle—the idea that when you are in the immediate vicinity there is no way to tell apart a freely falling system from an inertial system.

One of the surprising consequences of the equivalence principle is the idea that light bends in the presence of gravity. We can see how that works by imagining a glass elevator freely falling toward the ground at a high velocity close to the speed of light. Suppose a laser halfway up one of its walls is aimed at a spot halfway up the elevator's opposite wall. If you were inside the elevator, from that perspective, which is locally indistinguishable from being at rest, the laser would seem to send its beam straight across. However, if someone was standing on the ground watching the elevator, when the elevator was up high they'd see the light leave the laser halfway up one wall, gradually descend in space as the elevator rapidly falls, and hit the halfway point on the opposite wall at a point significantly lower than the starting point. Thus, the light would follow a parabolic trajectory, similar to the path a ball takes when rolled off a high ledge. (Horizontal constant motion, combined with vertical, downward acceleration due to gravity, produces a parabolic path overall.) If they then ignored the elevator itself and focused solely on the laser beam, it would seem to bend under the influence of gravity.

That gave Einstein a brilliant idea. Imagine if the whole universe could be described as an intricate network of locally inertial frameworks—

something like an immense bank of freely falling glass elevators—each plunging under the gravitational influence of the mass and energy in its local region. From the perspective of each local framework, light would travel in a straight line. However, what is straight according to one framework might appear bent in others observing from a different perspective. To track the path of light through space, therefore, requires revising the coordinate systems from point to point to accommodate those differences. The result is a global map of light tracks that shows how light responds within each local patch of matter and energy. As Einstein showed in his masterful general theory of relativity, completed in 1915, the terrain of that map is shaped by the amount of mass and energy in each region. Like a heavy rock denting an otherwise flat bed of wet clay, according to Einstein's equation, matter and energy warp the fabric of space and time. Spatial curvature, in turn, alters the path of everything that moves through the universe. Thus, Einstein was able to explain in a natural way why the planets move around the sun—because its immense mass distorts space in its vicinity—thereby replacing Newton's unsatisfactory explanation of "invisible threads" of gravity.

In his seminal paper on general relativity, Einstein made several key experimental predictions. Among those was the prognostication that starlight would bend in the vicinity of the sun as a result of the sun's mass warping space. Naturally, Einstein knew he couldn't test his forecast during normal daytime hours (too bright to see stars) or at night (none of the starlight passes anywhere near the sun). However, a total solar eclipse would offer the perfect situation to test solar light bending.

During the war, few German scientists kept in touch with their British counterparts. Einstein, though, who considered himself a pacifist and an internationalist, was a rare exception. He and Eddington, who was also a pacifist and who was avidly interested in general relativity, maintained a close correspondence. Thus, it was natural for Eddington, along with Dyson, to organize the key astronomical missions in 1919 to the Southern Hemisphere—specifically Sobral, Brazil, and Príncipe, an island off the coast of West Africa—to test Einstein's

Eldwick School, which Fred Hoyle attended from ages nine to eleven, before enrolling in Bingley Grammar School. CREDIT: Photograph by Paul Halpern.

hypothesis. Einstein was delighted by the teams' results and indebted to Eddington for his persistence.

STARS AND ATOMS

While Einstein became world famous, Eddington shared a measure of that fame, especially in Britain. The success of the solar eclipse research helped Eddington launch a second career as a science popularizer, alongside his primary mission as a research scientist. He wrote a number of widely read books, including some of the first popular works in the English language describing Einstein's theories. He eventually was knighted—one of the few scientists, including notables such as Newton, to receive such an honor.

One of the young minds Eddington influenced was Fred Hoyle's. While in high school, he borrowed Eddington's 1927 book *Stars and Atoms* from Bingley Public Library. Based on lectures Eddington had

Sparable Lane, a footpath near Fred Hoyle's childhood home, where he first encountered the wonders of the starry dome. CREDIT: Photograph by Paul Halpern.

delivered in Oxford, it examines questions such as what goes on in the interior of stars and features his novel idea that the fusion of chemical elements generates stars' energy. Although Hoyle's first love at that time was chemistry, he had developed a passion for astronomy as well. Eddington's book beautifully showed how those interests could be combined in the study of stellar nucleosynthesis, an area in which Hoyle would become a true pioneer.

Hoyle traced his fascination with stars to his preteen days at Eldwick School. By that time, he had overcome, to a certain extent, his anxiety about school that the traumatic botany incident at the Mornington Road School had induced. Starting out behind because of his many months of truancy, he very quickly caught up with and surpassed the other pupils. He also began to play more freely with other boys and girls, including a nighttime game resembling hide-and-seek. One day his search for a hiding place took him to Sparable Lane, an unpaved road through a wooded area near his house so narrow that it was essentially a foot path. (Sparable, from "sparrow bill," means a headless, thin nail, which is what the lane resembled.) On one side of the lane was a wall overlooking the valley, and he and a friend climbed onto it. It was there he had an epiphany about the wonders of the universe.

As he recalled: "When on top of a wall that perfect starlit night, I seemed to be in contact with the sky instead of the earth, a sky powdered from horizon to horizon with thousands of points of light, which, on that particular dry, frosty night, were unusually bright. We were out there for perhaps an hour and a half, and, as time went on, I became more and more aware—awed, I suppose—of the heavens. By the time I arrived back . . . I had made a resolve . . . that I would find out what those things up there were."[15]

Although Eldwick was superior to Mornington Road, it still didn't meet the brilliant young thinker's needs, especially once he became motivated to study the mysteries of atoms and the stars. His classes, with a wide mix of students of different ages and abilities, didn't challenge him much. Given his strong interest in science, Hoyle's parents hoped to find him a place in an academic high school that would prepare him well for university. The closest such school was Bingley Grammar School on the opposite side of town. To attend on scholarship, Hoyle needed to take a highly competitive qualifying examination.

When faced with the tough exam, however, some of his anxious feelings returned. Working painstakingly slowly, he skipped some of the math questions and muddled through part of the English language and comprehension section. Of the math questions he did answer, he indeed solved all of them correctly. So, when he got home from the test, he wasn't sure whether he had passed or failed.

Bizarrely enough, he ended up failing *and* passing. The first letter he received told him he hadn't made the cut. He was crushed. However, a scandal soon erupted when parents found out that the passing rate in the Bingley area was much lower than usual that year. Seats at the grammar school would be empty if they took the exam results at face value.

After much complaint, the headmaster, Alan Smailes, who had trained in mathematics at Cambridge, called Hoyle to the school for a private interview and a discussion with the head of chemistry. Hoyle must have impressed them both because they soon awarded him a scholarship. So off to Bingley Grammar School it was.

Hoyle felt lucky to have received the scholarship. Although his education remained largely self-motivated, rather than teacher-inspired, he began to feel even more confident about his studies. As the years flew by in high school, and he rose to the top of his class, he grew increasingly certain that he wanted to pursue a career in chemistry. It would be straightforward to do so, he thought. He had very high marks by then. All he needed was to qualify for a scholarship from the West Riding of Yorkshire authorities, which would enable him to study chemistry at Leeds University, perhaps the best university for science in Yorkshire.[16]

In a strange reversal, Hoyle ended up passing *and* failing the second qualifying examination—this time for university rather than high school. His marks on the test, while far from perfect, would normally have been suitable, but the authorities decided to be especially strict that year to toughen admission standards to more closely meet national averages. Thus, though he came away from the exam fairly certain that he'd soon be off to Leeds, fate pulled a nasty trick on him.

But even a field of ugly weeds might contain a four-leaf clover. Seeing Hoyle's disappointment, Headmaster Smailes took him under his wing and resolved to help find him a university placement. Familiar with Cambridge, Smailes urged Hoyle to study the materials for its scholarship exams in chemistry, physics, and mathematics. At first, Hoyle felt those problem sets to be overwhelming, but Smailes patiently coached him. With his heart still set on Leeds, Hoyle followed Smailes's advice to explore Cambridge and take the exams. It took several tries, but finally Hoyle was accepted at Emmanuel College, Cambridge, in the field of natural sciences. That's the college Smailes had attended as well. Hoyle put his dreams of Leeds and chemistry aside and, following in his mentor's footsteps, decided to complete the "Mathematics Tripos" program at Cambridge, which included theoretical physics along with mathematics.[17] He would start university there, at the age of eighteen, in autumn 1933.

It would be at Cambridge—where Eddington speculated about the properties of the universe, Dirac explored the nuances of the quantum world, and the researchers at Cavendish Laboratory, headed by Ernest Rutherford, probed the properties of the atomic nucleus (James Chadwick discovered the neutron there in 1932)—that Hoyle would make his mark, mainly in the subjects of astrophysics and cosmology. Hoyle's childhood, which began in the year the general theory of relativity appeared, ended with him on the road to delving into some of its monumental cosmic implications.

CRAFTING A UNIVERSE

Cambridge has a venerable connection with the study of the universe. There, back in the late seventeenth century, Newton had crafted his concept of the cosmos as a clockwork entity, created and set into motion by God. Newton had supposed that, after Genesis brought Earth and the heavens into being, the starry patterns of the sky would remain more or less the same until the end of time. Newton's laws of motion and gravitation—the core of classical physics—mandated that the celestial bodies function like machinery in their movements. Newton, who was a religious Christian, believed that only divine will might alter their endless rhythms.

In 1917, soon after he completed the theory of general relativity, Einstein decided to apply its formalism to the study of the universe. He didn't expect revolutionary results. Rather, he hoped to reproduce the stable Newtonian universe with one key difference: instead of absolute space and absolute time used to justify when objects are in the state of inertia (either at rest or moving with a uniform velocity), he aspired to apply a notion called Mach's principle, proposed by Austrian thinker Ernst Mach, which stated that inertia was set by the collective influence of the distant stars. Mach's concept of inertia seemed more tangible to Einstein than Newton's invisible, universal yardsticks and timepieces. Thus, Einstein's ardent hope was that his theory would reproduce the static, Newtonian universe, but in a natural, concrete, self-consistent way in the spirit of Mach.

Models of the universe were much simpler in the era when Einstein began to explore cosmology because they lacked the concepts of galaxies and clusters of galaxies beyond the Milky Way. Astronomers thought that spiral structures in the sky, Andromeda, for example, now known to be distant galaxies, each full of billions of stars, were simply nebulae (gas clouds) within the Milky Way. Therefore, without considering the existence of other galaxies, Einstein just needed a way to support a stable, relatively uniform haze of stars.

To solve his equations of general relativity, Einstein made some simplifying assumptions. One assumption is that the universe, at any given time, is homogeneous. Just like nuts and currants mixed well into a bowl of porridge, the mixture of contents of any given portion of the cosmos—stars, gases, and so forth—should look roughly the same as in any other. Correspondingly, because in general relativity the distribution of matter and energy sets space's geometry, its curvature in any given region should be roughly the same as that of any other. That is, it should be as smooth and regular as a marble ledge.

Also, Einstein assumed spatial isotropy, an even stronger condition than homogeneity. From Earth, looking out into space, astronomers count roughly the same number of stars in all directions. That implies all directions have approximately the same material distribution. From general relativity, that means spatial geometries must have the same curvature along any direction from Earth. Combining isotropy with homogeneity constrains space to look the same in all directions at any point—something like an infinite sheet of paper or a perfectly spherical ball.

In fact, there are only three types of spaces that are completely isotropic and homogeneous. The first is a hyperplane, the three-dimensional generalization of an endless, perfectly flat plane. One might imagine it as an infinite box. Mathematicians refer to it as having zero curvature.

The second is a hypersphere, the three-dimensional equivalent of the surface of a sphere. In the language of non-Euclidean geometry (the mathematics describing curved spaces), it has positive curvature. Imagine all of space curled up like a globe so that it had no real boundaries, just immense cycles in all directions. Just as on Earth, if you fly eastward long enough, you circumnavigate the globe and end up back at your starting point, the same would happen to a spaceship if it could possibly travel that long in a hyperspherical cosmos; the craft would circumnavigate space and return home. (Of course, such a feat far surpasses the power of any propulsion system conceivable today.)

Finally, the third possibility is a hyperboloid, or space of negative curvature—the higher-dimensional equivalent of a saddle shape. Just as a saddle bends downward in some directions (think of the sides hanging down from a horse) and upward in others (think of the front and rear, angled upward for support), a hyperbolic space curves differently along perpendicular axes. A perfect potato chip, curled two different ways along two different directions, has similar features. Like a flat (hyperplane) geometry, a hyperboloid is unbounded. An astronaut setting off in any direction would never return to her or his starting point. Yet another way of referring to hyperbolic space is "open" (endless), as opposed to hyperspherical space, which is "closed" (bounded). Flat, the third option, is similarly endless but, having zero curvature, represents a special case.

One might wonder, if space is curved, into what does it actually protrude? There are two ways to answer that question. The first is to postulate an unseen extra dimension (beyond the three dimensions of ordinary space and the dimension of time) that exists purely to allow space to be curvy, just like Earth's subterranean core, mantle, and crust support its round surface. Before scientists knew much about Earth's interior, they could infer the existence of a core (at least geometrically) merely because they had determined that its exterior must curve around something. Even if Earth was made of a material so strong it would be impossible to drill into, we could still make such an inference. Similarly, we might suppose that space bends around an inaccessible realm.

Physicists traditionally don't like to add experimentally unverifiable features to their models, however. (Arguably, that tradition has fallen by the wayside in recent years.) Having an extra dimension that is off-limits seems unphysical. Fortunately, there is another way of looking at curved, non-Euclidean geometries that does not require an extra dimension. As shown by a number of nineteenth-century mathematicians such as János Bolyai, Carl Gauss, and Nikolai Lobachevsky, one might simply relax the ordinary rules of planar geometry and make them more malleable. Whereas Euclidean geometry, the standard learned in high school, has strict guidelines—such as the sum of the angles inside a triangle is always 180 degrees, and that in a square, always 360 degrees—non-Euclidean geometry eschews such constraints, leading to more flexibility.

Hence, according to non-Euclidean geometry, if in a certain region of space, triangles contain more than 180 degrees (imagine forming a triangle using a portion of the equator and two lines of longitude that meet at the North Pole), we know it has positive curvature (theoretically bulging "outward"). Sectors of space with triangles containing fewer than 180 degrees have negative curvature. Finally, regions with triangles containing exactly 180 degrees are flat.

With three isotropic, homogeneous types of spaces to choose from, for his first cosmological model Einstein selected the case of positive curvature. To establish the conditions satisfying Mach's principle—in which inertia represents the combined influence of distant stars—he needed a finite universe because an infinite number of distant stars would wrongly produce an infinite effect. Since Einstein wanted to assume an eternal, static universe, he envisioned space as a hypersphere. That way, space was finite but, like a globe, had no physical boundaries.

MATTER WITHOUT MOTION OR MOTION WITHOUT MATTER

With all the geometric constraints determined, Einstein was ready to solve his equation of general relativity linking space and matter. In his formulation, he combined a hyperspherical geometry with a uniform stew of mass roughly approximating a universe sprinkled evenly with stars. Then he looked for viable mathematical solutions that were stable, not dynamic — showing how, as he and his contemporaries believed, space remains constant over time.

Suddenly, he realized that the equation of general relativity had a major problem. Mathematically, the cosmologies he found as solutions either expanded or collapsed, but wouldn't stay put in a stable configuration. Preferring a static universe to match the scientific consensus of his day that, on average, the cosmos does not change overall throughout the eons, he knew that wouldn't do. Rather than contemplate a dynamic universe, his predilection toward stability kept him focused on trying to make it work in the equation.

Finally, Einstein tried a last resort: adding a stabilizing term to the geometric side of the general relativistic equation. He called it the *cosmological constant*, symbolized by the Greek letter lambda, λ. He thus replaced the two-way connection between matter and geometry with a three-way connection that included the cosmological constant. To his relief, he found seemingly stable solutions, with positive curvature, a finite distribution of matter, and the cosmological constant. Although this method was as artificial and ugly as supporting an elegant building with scaffolding, at least it maintained the Newtonian idea of a static universe, while aiming for a Machian interpretation of inertia.

Admittedly, Einstein was uncertain from the start about the cosmological constant term. Accustomed to motivating everything he did with clear physical principles, arbitrarily inserting an extra term seemed odd. He sent a copy of his paper to Dutch physicist Willem de Sitter, characterizing his own invention as "outlandish." Around the same time, he wrote to his good friend, physicist Paul Ehrenfest, also in Holland, informing him that he expected to be certified insane and committed to an asylum for what he did to gravitational theory.

De Sitter immediately set out to explore the implications of Einstein's new rendition of general relativity as modified with a cosmological constant. He played with the model like a child with a new construction kit, seeing if he could find solutions other than the static universe model Einstein had concocted. There is only one real physical universe, so, given the reasonable constraints on space and matter that Einstein imposed, ideally there should be one unique solution. If there were more, then researchers would have to find other reasons why the universe behaves the way it does.

Much to his amazement, de Sitter did indeed discover a second, distinct solution. Instead of being filled with a smooth gruel of matter, as Einstein had projected, de Sitter could solve the equation by supposing the universe possessed no mass or energy at all. Rather, its contents consisted of absolute nothingness.

De Sitter left out the matter and energy to test within the context of general relativity the logical consequences of Mach's principle, which stated the contents of the universe produce its dynamics or its stability. If matter is needed to establish inertia, the lack of matter should produce a stable universe with no inertia. If the substance of the universe sets its dynamics, a lack of substance should lead to no dynamics. However, that's not how de Sitter's solution behaved. Rather, it showed a universe that was expanding exponentially. In fact, the nothingness drove the most virulent kind of growth.

De Sitter's alternative model was not good news for Einstein, who had hoped that the cosmological constant would ensure stability. Yet it was clear that it still allowed for out-of-control solutions. And a comment de Sitter made at the end of one of his published articles on the topic surely hit hard: "It cannot be denied that the introduction of this constant detracts from the symmetry and elegance of Einstein's original theory, one of whose chief attractions was that it explained so much without introducing any new hypothesis or empirical constant."

Einstein had begun his venture into gravitation in part to make the static Newtonian universe more physically intuitive, in line with Mach's emphasis on the tangible. Yet as de Sitter's interpretation of his first cosmological work showed, it was clear that he had missed the mark.

Einstein had stumbled into an occupational hazard of being a maverick theoretician. By basing a revised version of general relativity on his personal intuition about the universe, he had bet that his impulses were so logical that they must be correct. He had imagined the cosmos as a kind of stable tent, supported by its own internal structure and the combined influences of the stars. When that idea didn't work, rather than considering alternatives, he had persisted in trying to prop it up with the cosmological constant to maintain its stability.

In truth, we know now that space does indeed expand over time. Einstein's desire to keep the universe static turned out to be the real mistake. It would take the work of astronomer Edwin Hubble in 1929 (anticipated by the work of Vesto Slipher and others), who observed that distant galaxies, "island universes" in their own right, were receding from ours, and receded faster the more distant they were, for Einstein finally to budge on that issue, admit that the universe is dynamic after all, and drop the cosmological constant.

In the 1920s, even before Hubble's observations, theorists Alexander Friedmann and Georges Lemaître introduced dynamical models of the universe that did include matter and thus were more physically realistic than de Sitter's model. While clinging to the notion of stability, Einstein had dismissed these models—only later to regret his mistake when he saw the light about cosmic growth. After Hubble's measurements suggested an expanding universe, an assortment of additional models for cosmic growth emerged in the 1930s, including several in which Einstein was himself involved.

Although Einstein's intuition about a static universe proved wrong, the cosmological constant was not forgotten. When Hoyle and his colleagues Hermann Bondi and Thomas Gold proposed the steady-state model of the universe, they kept de Sitter's model of an empty universe with a cosmological constant in mind. Separating himself from Bondi and Gold, Hoyle elected to replace the constant with a matter-generating entity called a creation field. The dynamics were similar, nonetheless, to the effect of a cosmological constant.

More significantly, Einstein's discarded term would be resurrected in the late 1990s when astronomers discovered that not only is the universe expanding but also its expansion is accelerating. The cosmological constant seems to model that extra boost. Although that was hardly Einstein's intention when he introduced it, sometimes old ideas get recycled for new purposes.

Einstein was certainly neither the first nor the last brilliant scientist to operate on the basis of hunches. In that sense, both Gamow and Hoyle followed in his footsteps, coming up with extraordinary ideas based on splendid intuition. In the late 1940s, each strived to put flesh on the skeletal models of cosmic expansion, developed by earlier thinkers, by lending their own intuitive notions about the physical universe. The result was the competing descriptions of the expanding cosmos: the Big Bang and steady-state theories.

As maverick, intuitive thinkers similar to Einstein, Gamow and Hoyle would also sometimes fall short in their conceptions. Typically, Gamow would then drop the idea or hand it off to others for improvement. Stubborn and independent, Hoyle would persist anyway, even when his hunches could not be supported by credible evidence. Throughout his later years, he'd advocate various renditions of the steady-state universe, long after mounting evidence seemed to disprove the concept. Basing a scientific model on hunches about truth and beauty is a tricky business. Impulsive ideas might prove spectacularly correct or remarkably wrong.

CHAPTER TWO

Preparing the Battlefield

ANTICIPATIONS OF A COSMOLOGICAL CLASH

> If one considers a physically bounded volume, particles of matter
> will be continually leaving it. For the density to remain constant,
> new particles of matter must be continually formed within that
> volume from space.
>
> —ALBERT EINSTEIN, "*Zum kosmologischen* Problem,"
> translated by Cormac O'Raifeartaigh and Brendan McCann

W HEN FRED HOYLE, AS A BOY, MARVELED AT THE STARS OVER
Gilstead, and George Gamow aimed his childhood telescope
at the skies over Odessa, no one had any idea that, within a short time,
humanity's concept of space would expand radically. We would soon
come to recognize that the universe is far more immense than we once
believed, that it is full of galaxies, and moreover, that the bulk of those
galaxies are moving away from ours, an indication that space is expanding. Not even Einstein, wrapping up his brilliant, novel explanation
of gravitation, anticipated that it would be used to model such a revolution in cosmology. Even when de Sitter pointed out how a dynamic
universe might be possible, both he and Einstein saw that solution as a

mathematical quirk rather than as a serious alternative to the traditional notion of a static universe. But a wave of key observations—by Vesto Slipher, Henrietta Leavitt, Edwin Hubble, and others—would soon wash over the sandcastle visions of a static cosmos. Unmistakable evidence of cosmic growth eventually inspired competing interpretations: the Big Bang, proposed by Georges Lemaître and developed significantly by Gamow to explain how matter is created in the universe, and the steady-state, proposed by Hoyle, Bondi, and Gold.

In 1915, the year when Hoyle and the general theory of relativity were born, American astronomer Vesto Slipher of Lowell Observatory in Flagstaff, Arizona, noticed a curious thing about some of the pinwheel-shaped objects in the heavens called spiral nebulae. *Nebulae* (singular: *nebula*) means "gas clouds." Before observational evidence began to suggest otherwise, the majority of astronomers presumed nebulae were the hazy incubators of stars within the Milky Way. Using a spectroscopy method that he had developed, Slipher determined the radial velocities (the rate at which the nebula was speeding toward or away from us) of fifteen of these spiral bodies and found that almost all of them were moving away from us at great speeds; only a handful, including Andromeda, was moving toward us. Perplexingly, the nebulae in Slipher's survey moved twenty-five times faster, on average, than typical stars in the Milky Way.

Slipher's Doppler shift radial velocity measurement technique, still used today, involves using a spectrograph to break up the light from an astral object into its spectral line components, and then analyzing the positions of those lines relative to the standard pattern for various elements found in stars. Normally, the emission spectrum of an element is a rainbow pattern of the frequencies (colors, in the case of visible light) of light emitted when electrons jump from one state to another in the atoms making up the gaseous state of that element. Correspondingly, the absorption spectrum describes the pattern of the frequencies of absorbed light. According to quantum mechanics, such changes in energy states happen regularly, with the probabilities determined by special rules that show how each element's atoms' characteristic array of energy levels dictates the types of movements the atoms can make. Therefore, an element's spectral pattern is a predictable arrangement of frequencies—something like using the same bar code for identical merchandise in a store.

Say, for instance, the gas is hydrogen. The electrons in hydrogen atoms have a fixed ladder of energy states they can occupy, as determined by quantum mechanics. Therefore, electrons can drop down or leap up only certain quanta (discrete amounts) of energy, emitting or absorbing photons (particles of light) of particular frequencies in the process. Following a quantum rule, each photon's frequency depends on its energy: the more energetic the photon, the higher its frequency. That produces the standard emission, or absorption spectral pattern.

However, physicist Christian Doppler showed that, for any given collection of waves emitted from a source, movement of the waves toward an observer appears to shift their pattern of frequencies toward the higher end of the spectrum and movement away from an observer appears to shift the waves' pattern of frequencies toward the lower end. In the case of sound, inward motion toward the observer causes a Doppler shift toward a higher pitch, and outward motion away from the observer shifts the sound waves toward a lower pitch. We hear the Doppler effect when a fire engine races toward us and its siren sounds like a high-pitched wail. Then it speeds away, and the same siren sounds like a low-pitched rumble.

Doppler examined sound specifically, but the effect he described also holds true for light. If a luminous source, such as a star, is moving toward us, its light spectrum shifts toward the higher, or bluer, end of the spectrum. That is called a "blueshift." In contrast, if a source of light is speeding away from us, its frequency pattern shifts toward the lower end of the spectrum, which is called a "redshift." Slipher recognized that the method could be used to pin down the inward or outward speeds of distant astral objects, such as the spiral nebulae he observed. Yet, once he assembled the data, he couldn't adequately explain the huge radial velocities he found and why most nebulae were moving away from us.

A minority theory at the time of Slipher's observations was that the spiral nebulae were other galaxies, or "island universes," well beyond the boundaries of the Milky Way. Thinkers such as Immanuel Kant and Edgar Allan Poe (in his final work "Eureka") had speculated such, but there was no proof. Such evidence was soon forthcoming, showing that Slipher's results were early evidence that the vast majority of other galaxies are moving away from ours—suggesting that the universe is expanding. Hoyle and others would later argue that Slipher didn't get enough credit for his discovery.

Ironically, the means of testing the distances, as well as the velocities, of the spiral nebulae was also close at hand. In 1912, Henrietta Leavitt of Harvard Observatory discovered a remarkable relationship between the periods (time between regular pulses of stellar radiation) and the brightness of cepheids, a type of variable star. She was looking not at the nebulae but rather at objects in the Large and Small Magellanic Clouds—sky features that astronomers ultimately would reveal to be dwarf galaxies in the "Local Group" of the Milky Way. Her formula, when generally applied, would allow cepheids to be used as "standard candles": objects of known absolute brightness, as reliable as the labels on 60-watt light bulbs in predicting their power output. And once you know how intrinsically bright a bulb is, you can figure out how far away it is in an otherwise dark hallway: first, you measure how bright it looks, compare it to how bright it actually is, and then—because light intensity decreases a standard amount over a certain distance—you know its distance away from you. But Leavitt, who was a "computer" (a term used for people who did arduous calculations), not a professor, had little say in choosing her next project and was asked to move on to another assignment. Sadly, she died in 1921 at the age of fifty-three, before witnessing the fruit of her immensely powerful formula, now known as Leavitt's law.

In 1920, the Great Debate (as it would later be called) between two eminent astronomers, Harlow Shapley and Heber Curtis, about the size

of the universe took place at the recently constructed Natural History
Building (now the National Museum of Natural History) of the Smithsonian Institution in Washington, DC. Shapley argued that the spiral
nebulae were distant denizens of the Milky Way, which he projected to
be much bigger than previously believed. Curtis asserted, on the contrary, that the spiral nebulae were galaxies similar to the Milky Way,
but well beyond its boundaries. The universe, he contended, is full of
spiral galaxies, including the Milky Way, Andromeda, and many others.
To bolster his position, Curtis maintained that the Milky Way must be
relatively small. To make a long story short, Shapley was right that the
Milky Way is enormous and Curtis was correct in that it is but one of
many galaxies. The universe turned out to be far greater than either of
them and their contemporaries imagined.

To describe a vast, expanding universe would require cosmological models that accounted for both matter and motion. Thus, neither
Einstein's nor de Sitter's 1917 models would make the cut. Fortunately,
general relativity held a plethora of surprises, some of which matched
the burgeoning observational evidence. In Saint Petersburg, Alexander
Friedmann, a clever mathematician who would instruct George Gamow, began to unpack such dynamic solutions of Einstein's equation.

THE SAVANT OF SAINT PETERSBURG

In 1922, when eighteen-year-old George Gamow arrived in Saint Petersburg, he was almost completely alone in a city far different from the
southern port town he had come from. The city had only recently lost
its status as the capital of a vast empire, and Moscow now governed the
newly formed Soviet Union. As had happened in other formerly imperial
cities after revolutionary change, the wealth drained from the palaces
and other lavish buildings, leaving them like bloodless corpses. Nevertheless, the university persisted in delivering high-quality education.

Before he started university, Gamow met up with a friend of his father, V. N. Obolensky, professor of meteorology at the Forestry Institute.

Obolensky offered Gamow one of the few practical jobs he would hold in his life—recording weather data, such as air pressure and wind velocity, at a meteorological station. After some time taking measurements, he was ready to do something more creative. Once he began his university studies, his thoughts turned to Einstein's theories. That naturally led him to consider working with Friedmann. As Gamow recalled:

> At this time I was mostly interested in relativity. . . . Doing relativity there was Professor Friedmann, the man who has shown that Einstein is wrong and that his cosmological equation has a time-dependent solution. . . . And I actually thought I would work with Friedmann. . . . He was a professor of mathematics, but he was interested in the application of mathematics, and what he was mostly doing was hydrogen aerodynamics. . . . He was also interested in mathematics of relativity, and then he found this mistake of Einstein and reported what is now known as Friedmann universe. . . . He was giving a course on relativity—the first formal course on mathematical relativity I took was from him.[1]

Gamow's relativity course with Friedmann would prove life-changing. He began to revel in the joys of theoretical physics, which sometimes predicted unexpected oddities about the universe—clocks slowing down, rulers shortening. He was lucky to come of age during one of the most revolutionary eras of modern physics.

Friedmann, as Gamow would learn, had indeed taken time off from his aerodynamics research to explore the hidden possibilities of general relativity. Retracing the initial steps of Einstein and de Sitter, he examined the three different geometries that were isotropic (the same in all directions) and homogeneous (identical at all places). Recall that these are the positive-curvature hypersphere, the negative-curvature hyperboloid, and the zero-curvature hyperplane—three-dimensional generalizations of the surface of a sphere, the surface of a saddle, and the surface of a plane, respectively.

Russian physicist Alexander Friedmann,
a pioneer of dynamic cosmology.
CREDIT: AIP Emilio Segrè Visual Archives.

Brilliantly, Friedmann then defined a "scale factor" that would represent the relative size of space over time. If the scale factor doubles, then points on the shape in question become twice as far away from each other. Imagine painting a series of evenly spaced dots on a basketball, a flexible saddle, and a sheet of rubber. Blow up the basketball, and the dots separate from each other uniformly. Similarly, stretch the saddle evenly, and pull the sheet equally from all sides at once, and the dots painted on them would move apart in a uniform way as well. Conversely, deflate the ball and shrink the saddle and sheet, and the dots on each would move closer and closer together.

With each shape category, Friedmann included a flexible scale factor to set up the geometric side of Einstein's equation of general relativity. That is, he let the type of geometry (positive curvature, negative curvature, or flat) determine the shape of space and the changeable scale factor determine its size. He built the other side of the equation, representing mass and energy, by inputting the density of matter. He also allowed that value to be flexible. Finally, he made the cosmological constant an option, which enabled him to consider what role its absence or presence might play.

In short, with an eye for aerodynamics, Friedmann had built a model of the universe that resembled a hot air balloon. Turn up the temperature of its furnace, and it expands. Crank the temperature down, and it contracts. Fueling the furnace is a representation of the density of matter in the universe.

To solve Einstein's equation and find the specific dynamics associated with each geometry, Friedmann needed to constrain the density term. For a positively curved space, the density needed to be larger than a certain critical value. For negative curvature, the density needed to be smaller than that value. Finally, a space of zero curvature must be driven by a density that exactly matched the critical value.

These three possibilities led Friedmann to find three distinct solutions. Leaving out the cosmological constant, he discovered that positive curvature, with its relatively high density (compared to the critical value), led to a scale factor that started small, grew until it reached a maximum value (corresponding to a maximum radius of the hypersphere), and then began to shrink down to a point. Thus, it represented a "closed universe," both in space and time, being finite in each. That scenario would later become known popularly as the "Big Crunch."

Negative curvature, on the other hand, with its relatively low density, produced a scale factor that kept growing forever, but gradually slowed over time. There would be no recontraction. Such an "open universe" scenario has been dubbed in modern times the "Big Whimper" (after a passage from T. S. Eliot's famous poem "The Hollow Men": "This is the way the world ends. Not with a bang but a whimper").

Finally, a universe represented by a space of zero curvature would forever teeter on the brink of collapse, but never do so, like a frail tightrope walker who continues to perform show after show with grace, knowing that one slip would end his career but luckily remaining balanced each and every time. Similarly, for a "flat universe," even the slightest excess of density would trigger eventual collapse. But if the density remains precisely at the critical value, space would expand forever.

You don't need to comprehend the higher mathematics of general relativity to understand the three scenarios. Imagine launching a robot-controlled rocket, hoping it will clear the atmosphere, head into space, and end up orbiting Earth as a satellite. On the first attempt, you weigh it down with a heavy cargo (representing an overly dense universe) and fire it off. It journeys up into the sky, slowing all the while. It releases some of its ballast to propel itself, but not enough. Laden with excess contents, it is unable to clear Earth's gravity. It reaches its highest point, and then falls back toward the earth, crashing into the ocean not far from the launch pad. That scenario represents the "closed universe" model, ending in a "Big Crunch."

You try again with a second rocket. This time you include much less cargo (representing an under-dense universe). As the rocket ascends, it jettisons most of its ballast. It enters deep space with enough momentum to keep going. Over time it slows down, but not enough to remain captured by gravity as a satellite. Rather than orbiting, therefore, it clears Earth's orbit and continues onward. That scenario represents the "open universe" model, ending in a "Big Whimper."

Finally, you make careful calculations about the weight of cargo needed, and try one last time with a third rocket. You launch it, and it climbs into space. Its ascent slows less than in the first scenario, but more than in the second. You and your team celebrate as you see it begin to orbit Earth. The critical cargo weight (standing in for the critical density of the universe) was just right, leading to the equivalent of the "flat universe" scenario.

Friedmann wrote up his results in a paper titled, *"Die Krümmung des Raumes"* ("The Curvature of Space"). It was published in the prestigious German journal *Zeitschrift für Physik*, which had an international audience (German was the lingua franca of science at that time, along with English and French). The article offered a guidepost in considering the possibilities of an evolving universe. But given that there was no evidence yet for an expanding universe (aside from Slipher's data, which had yet to be properly interpreted), most contemporary readers likely saw Friedmann's results as a purely mathematical exercise.

Einstein, however, who had a philosophical stake in the legitimacy of his support for a static universe, took the article as an affront. He responded to Friedmann's paper with a brief note in the *Zeitschrift* indicating that it "seemed suspect" because of a changing density of matter. He argued that constraints he had imposed on the original equation should make the density of matter constant and (taking the cosmological constant into account) the size of the universe constant as well. Hence, in Einstein's view, Friedmann simply had erred. Case closed.

Friedmann was dismayed that Einstein had entirely missed his point. He wanted to explore the richness of nonstationary solutions, including ones with negative curvature. He was not calling Einstein's cosmological model into doubt. Rather, he was suggesting a whole new gamut of interesting possibilities.

Friedmann wrote a letter to Einstein to make his points clearer and, if possible, to correct the record. He detailed his calculations carefully, pointing out to Einstein their validity, and calling for a response. "Do not, highly esteemed Professor," he pleaded, "deny me notification about whether my calculations discussed in the present letter are right."[2]

After some time during which he received no reply to his letter, Friedmann was exasperated. Having his paper continue to be seen as erroneous would be a needless blot on his record that Einstein, after a fresh look, could easily wipe away. Fortunately, one of Friedmann's colleagues at the university, Professor Yuri Krutnov, happened to be planning a rare trip to Berlin (journeys abroad required special permission in Bolshevik Russia). Friedmann asked Krutnov to meet with Einstein to explain the situation.

Thanks to Krutnov's intervention, Einstein realized that he had been mistaken to dismiss Friedmann's work so abruptly. He published a short note in the *Zeitschrift für Physik* explaining that he had misjudged

Friedmann's paper. "I consider Mr. Friedmann's results correct and illuminating,"[3] he wrote. According to Gamow, Einstein also conveyed his personal regrets to Friedmann, albeit in a "somewhat grumpy letter."[4]

THE LOST OPPORTUNITY

Gamow so enjoyed Friedmann's course "Mathematical Foundations of the Theory of Relativity" that he keenly hoped to work with the brilliant professor on research projects. Physics history might have been very different if Gamow had been able to follow through on this dream.

Astronomy was enjoying extraordinary success in establishing the scale and content of the universe, thanks to the work of Edwin Hubble at Mount Wilson Observatory in California. A gifted astronomer, Hubble was also quite a character. Born in rural Missouri, and adept in various sports—including baseball, basketball, running, and boxing—in his early life he seemed destined to be an athlete. While attending the University of Chicago, he led their champion basketball team to great success in 1907 and 1908. Although he had a passion for astronomy, he wasn't even studying science at that point. Rather, he was considering pursuing a law degree. Then, a Rhodes Scholarship to Oxford transformed his life. He dropped his Midwestern twang and began to affect a posh English accent, started smoking a pipe and wearing tweed jackets or capes. He also switched from pre-law to science. After returning to the United States in 1913, he taught physics, math, and Spanish at New Albany High School in Indiana for a year, while coaching its basketball team. He then went back to the University of Chicago and obtained a PhD in astronomy in 1917. Thanks to his experience at Chicago's Yerkes Observatory, he was recruited to Mount Wilson, where he would make his mark as a brilliant watcher of the skies.

Hubble set out to resolve the Great Debate by establishing the true distance of Andromeda. For several months in 1923, using the hundred-inch Hooker telescope, the largest in the world, he was able to image a number of features in that astral object, including what turned out to be a cepheid, a variable star. By comparing exposed photographic plates taken in succession, he plotted its "light curve" (how its apparent brightness changed over time). That offered him the star's period value, which using Leavitt's law generated its intrinsic brightness (power output at the source). Finally, by comparing the apparent brightness and intrinsic brightness of that standard candle, he was able to ascertain its distance. By the end of 1924, he had found a dozen cepheids, which enabled him to increase the precision of his distance estimate. He established, beyond the shadow of a doubt, that Andromeda lay far beyond the periphery of the Milky Way. Because of its remoteness, Hubble could estimate that it was much larger than once believed. It wasn't a nebula at all. Rather, it was a full-fledged galaxy—a sister of the Milky Way.

Given such revolutionary results, it would have been a great time for Gamow to join the field. Alas, his dream collaboration with Friedmann would never come to pass. In July 1925, donning his meteorological (rather than relativistic) hat, Friedmann set off on a record-setting balloon expedition into the skies over Russia. As his craft rose more than twenty thousand feet in the air, he collected weather data that he hoped would help him develop novel models of the thin atmosphere at that lofty altitude. Unfortunately, he was not prepared for such harsh conditions and developed a chill. Sometime later, perhaps to recover, he went on vacation in the Crimea. Upon returning home, he developed a fever and was diagnosed with typhoid. He was hospitalized but died two weeks later, on September 16, 1925, at the age of thirty-seven.

Gamow was devastated by the loss of his mentor. The physics department assigned him to work with Krutnov instead. Krutnov offered him a research project on how to apply the concept of quantized energy (energy that is transferred in discrete amounts) to the behavior of a pendulum. Compared to grappling with the very nature of space and time, Gamow found the substitute project exceedingly dull. To make matters worse, the change in supervisors meant that his university education would be extended by an extra year. He also had new course requirements to meet in Marxist theory, which the Bolshevik government had imposed as it became increasingly oppressive in its ideological demands. By then, following Lenin's death, the university had been renamed the University of Leningrad, in tandem with the change in the city's name.

In his frustration, Gamow found some diversions. He latched onto a small group of fellow theoretical physics students, including Lev Landau, nicknamed "Dau," and Dmitri Ivanenko, nicknamed "Dim." Gamow's nickname was "Geo." The three friends dubbed themselves the "Three Musketeers." Often, they were joined in their social events by two female student friends, Irina Sokolskya, who had a penchant for sketching caricatures (a skill Gamow himself would acquire), and Yevgenia Kanegeisser, who had a knack for poetry. At the same time, Gamow discovered the mind-altering properties of alcohol, particularly vodka. Unfortunately, he would develop a drinking habit that ultimately took a toll on his health. On a healthier note, he acquired an interest in playing tennis. And, like Hoyle, one of his fondest pastimes was going to the movies, where he'd enjoy the adventures of heroes of the big screen.

Gamow's filmgoing fed his nonconformist mindset. Cinematic heroes often rebel against the expectations of others. A gunslinger might take matters into his own hands to restore justice, even if that meant technically breaking the law. A stranded sailor might find a makeshift way to restore a damaged craft. For such rebels, personal freedom is to be preserved at all costs. In Hollywood movies, protagonists often gravitate, therefore, toward lands beyond staid civilization—the open sea, remote desert islands, the Wild West, outer space, and other frontiers. Traditionally, America has represented a dreamscape of such possibilities—its stark vistas and seemingly endless roads open to exploration by hiking, motorcycling, or other means. Gamow shared with Hoyle, and so many others of their times, that romantic view of rebellion, escape, and a yearning for the frontier.

The Three Musketeers stimulated each other's way of thinking in much the same way as Einstein's group of intellectual friends, called the "Olympia Academy," had motivated him in the development of relativity. Gamow and the other Musketeers explored the implications of the newly developed quantum mechanics and collaborated on several research papers on the topic. Later in life, Landau became a highly prolific Soviet physicist, made seminal contributions to low-temperature physics, and was awarded the Nobel Prize in 1962 for his explanation of the superfluidity of liquid helium, a phenomenon in which fluid helium cooled to a certain temperature flows without friction. Ivanenko would make important contributions to physics as well, particularly in nuclear physics and particle theory.

THE PRIMEVAL ATOM

The tragic death of his mentor was a horrific loss, but Gamow had been lucky to have a course in relativity, a subject rarely taught in the mid-1920s, that would prove incredibly useful for his career. Relativity had not yet entered the canon. Physics curricula were grounded almost exclusively in the classical approach of Newton and Maxwell, with late-nineteenth-century thermal physics being the most modern subject generally taught.

Globally, perhaps the most eminent professor of the subject at the time, aside from Einstein himself, was Arthur Eddington, a fellow at Trinity College, Cambridge. With the "street cred" of having been one of the organizers of the eclipse expeditions to test general relativity's predictions, a keen interest in the philosophical aspects of physics, an openness to novel ideas, and a knack for writing, Eddington was the perfect expositor of Einstein's work. His popular works included *Stars and Atoms*, one of the books that inspired Hoyle; *Space, Time, and Gravitation*, one of the first popular books about general relativity; and *The Mathematical Theory of Relativity*, one of the first textbooks on the subject; and numerous other writings.

Belgian scientist and priest Georges Lemaître, who proposed that the universe once was extraordinarily dense. CREDIT: Photograph by Dorothy Davis Locanthi, courtesy of AIP Emilio Segrè Visual Archives, Locanthi Collection.

One of Eddington's most successful protégés in the field of general relativity was Georges Lemaître. Born in Charleroi, Belgium, in 1894, Lemaître had a talent and passion for math and physics, along with a deep devotion to his Catholic faith. A man of great humility and honesty, he saw his scientific and religious interests as strictly separate, but wholly compatible. In 1920, he graduated from the Catholic University of Louvain with a doctorate in physics and mathematics and then joined a seminary to train for the priesthood. In 1923, after he was ordained, he was offered the opportunity to further his studies at the University of Cambridge, where he spent a year under the guidance of Eddington. With Eddington's training, he became a prodigious disciple of relativistic thinking. He then spent a year at Harvard Observatory working with Shapley. In 1927, he earned a PhD from MIT based on calculations in general relativity and cosmology

Clock Tower, Great Court of Trinity College, Cambridge. Credit: Photograph by Paul Halpern.

that drew from his experience with Eddington, then returned to Louvain to assume an appointment as professor of physics.

The same year, Lemaître published a remarkable paper, *"Un Univers homogène de masse constante et de rayon croissant rendant compte de la vitesse radiale des nébuleuses extragalactiques"* ("A Homogeneous Universe with Constant Mass and Increasing Radius Explaining the Radial Velocity of Extragalactic Nebulae"). This work was the first to draw upon observational data to argue for an expanding universe. Lemaître combined the Doppler shift results of Slipher, which detailed the enormous outward velocities of many spiral nebulae, with the findings of Hubble, which demonstrated that at least some of the nebulae, likely galaxies, lay well beyond the Milky Way, to deduce that distant galaxies are receding (moving away from us).

The logical conclusion, Lemaître argued, is that space is expanding. It was not true that Einstein's matter without motion and de Sitter's motion without matter were the only options. Matter with motion was a distinct third possibility. Lemaître then, in a manner similar to Friedmann (whose work he wasn't yet aware of), showed precisely how the general theory of relativity predicted an expanding universe for a ball filled with matter. He imagined that the ball started at a fixed radius—that is, it was temporarily stationary—and then expanded, like an otherwise stable pufferfish that starts to inflate upon sensing predators. Because of that key paper, many historians consider Lemaître the "Father of the Big Bang."

Why didn't Lemaître's bold assertion of an expanding universe generate headlines around the world? For one thing, he had published his paper, written in French, in a Belgian journal that was scarcely read by scientists in other countries. Moreover, Lemaître was a quiet, humble cleric, with little desire for publicity. He relished sharing his idea with scientists familiar with general relativity—such as Einstein, of course—but was not prone to making his case in the media.

Einstein happened to be in Belgium around that time for the famous 1927 Solvay Conference, which is notable for his heated debate with Bohr about the interpretation of quantum mechanics. Lemaître used the occasion to chat with Einstein about his model. Einstein, in turn, informed the young theoretician about Friedmann's earlier models and remained dubious about an expanding universe.

Hubble, meanwhile, continued to employ the Hooker telescope at Mount Wilson to take distance measurements of other spiral galaxies (still often referred to as "extragalactic nebulae") using the standard candle technique of comparing the apparent and intrinsic brightness levels of cepheids. When he combined his data with Slipher's radial velocity results, he noticed a distinct pattern. Setting aside the closest galaxies, such as Andromeda and the Large and Small Magellanic Clouds, which along with the Milky Way form the Local Group, all other galaxies in space are receding from ours at a rate that increases with distance. In

American astronomer Edwin Hubble, sitting in front
of the Hooker Telescope. Hubble discovered evidence
for the recession of galaxies, which Lemaître and others
interpreted as indicating that the universe is expanding.
CREDIT: Hale Observatories, courtesy AIP Emilio Segrè
Visual Archives.

other words, the farther the galaxy, the faster it is receding. He calculated
the proportionality between speed and distance, a ratio now known as
the Hubble constant. He announced his results to a stunned astronomical community in 1929.

Lemaître had found the same effect using earlier, sparser data.
However, because it had been published in a Belgian journal, the international community remained largely unfamiliar with his result. To
offer him long-overdue credit, in 2018 the International Astronomical
Union voted to change the name of the principle that states the recession of galaxies is proportional to their distance from "Hubble's law" to
the "Hubble-Lemaître law."

Immediately after the publication of Hubble's paper, the global astronomical community began to shift toward accepting an expanding
universe. Growth of the cosmos naturally explains the fact that farther
galaxies are moving faster. One can picture the situation by imagining
a concert in a stadium, with the audience sitting around a circular stage
in rows spaced three feet apart. Suppose a safety inspector suddenly de-
cides that audience members must instead be seated twice as far apart;
that is, rows must be separated by six feet. Once the announcement is
made, the first row would move back an extra three feet to a row a total
of six feet away from the stage. The second row, however, must move
back six feet to sit in a row a total of twelve feet from the stage; the third
row moves back nine feet to a row a total of eighteen feet from the stage,
and so forth. If the movements all happened at once, those in the second row would have to move twice as fast as those in the first row; those
in the third row must move three times as fast as those in the first row,
and so forth. Therefore, in an expanding stadium seat arrangement, outward speed is proportional to distance from the stage. The same thing
happens in an expanding universe. Therefore, the Hubble constant is a
measure of the expansion rate.

Hubble didn't advocate that leading interpretation himself, though.
As astronomer Allan Sandage, in many ways Hubble's disciple who had
worked closely with the man at various observatories in California, ac-
knowledged, Hubble was reluctant to commit to the expanding universe
scenario without ruling out every other possibility. He believed that his
role was simply to present the analyzed data and leave the interpretation
to others.

Hubble "clearly wanted to sit on both sides of the fence," recalled
Sandage. "He clearly says 'Well, the true expansion explanation has a
lot of difficulties to it, and if there's not true expansion but some unknown law of physics then the relation should be this way, and if it's true
expansion then they should be another way,' and he always discussed
those two possibilities."[5]

Eddington urged Lemaître to submit his 1927 paper, translated into English, for publication in the *Monthly Notices of the Royal Astronomical Society* so that it would reach a wider audience. Lemaître agreed, and an English-language version was published in 1931. Strangely, however, a few sections of the original were omitted in the translation, including his earlier estimate of what came to be known as the Hubble constant.

For many years, the omission was a mystery. The excluded passages presented clear evidence that Lemaître was not only the first to predict an expanding universe based on observational data but also the first to develop Hubble's law and Hubble's constant. Could the journal's editor have left certain items out to bolster Hubble's claims?

In 2011, astronomer and science writer Mario Livio solved the riddle of the missing text.[6] Livio uncovered a revealing letter from Lemaître to the editor. It turns out that Lemaître, who had translated his paper himself, had left out certain findings because he preferred not to distract readers from Hubble's results, which were based on more extensive data. In his supreme modesty, Lemaître was little interested in what history would think of him. He only wanted the most accurate estimates to be available—which he believed were Hubble's, not his.

At the time, Lemaître was also motivated to deemphasize his 1927 paper in favor of his new idea about the origin of the universe. Instead of it having started out large, with a finite radius, and then growing even bigger from there, he proposed that it began as a super-dense ball of matter that had expanded from a point into its present-day size, like a living creature starting out as an embryo. He called his idea the "primeval atom" hypothesis, also known as the "cosmic egg."

Lemaître unveiled his notion of a universal creation at the centenary meeting of the British Association for the Advancement of Science, held in London in September 1931, where de Sitter and other eminent astronomers were present. Regarding an absolute beginning of time, he found himself at odds with Eddington and many other astronomers, who found the notion of a single instant of creation more akin to theology than to science. Although Lemaître himself was always careful to distinguish the scientific from the religious, the fact that a priest was suggesting a kind of "Genesis moment" may have served to alienate those who were suspicious of organized faith. Indeed, disdain for the notion of a beginning helped drive interest in the steady-state model later on.

ANTICIPATIONS OF STEADY-STATE

On January 29, 1931, during a research trip to Southern California to consult with scientists at Caltech, Einstein had the opportunity to visit Mount Wilson Observatory and meet with Hubble. Hubble was delighted, no doubt, to show Einstein the enormous telescope with which he had made his groundbreaking discoveries. By then, Einstein had changed his mind about the universe being static. Unlike Hubble, he became a zealous convert to the belief in an expanding cosmos.

Einstein would later discuss with Gamow how he believed that adding the cosmological constant in 1917 had been a mistake. If he had just left it out, he would have arrived at the solutions of Friedmann and Lemaître and predicted an expanding universe. Gamow recalled Einstein calling it "the biggest blunder he ever made in his life."[7] (Note that Linus Pauling once reported that Einstein told him that his letter to President Franklin Roosevelt about atomic bomb research was "the one great mistake in his life," so such recollections must be taken as anecdotes, not established fact.)

Sparked by Hubble's observations and his newfound appreciation of the work of Friedmann, Lemaître, and de Sitter, Einstein began to craft his own expanding cosmologies. His tinkering was part of his stubbornness. He felt driven to find the most elegant solution to any problem, the one that seemed most natural.

Einstein's first published expanding universe model (nicknamed the "Friedmann-Einstein universe"), submitted for publication in April 1931, dropped the cosmological constant and reverted to his original theory of general relativity. It essentially employs Hubble's constant as a means of setting the time scale and other parameters relevant to Friedmann's positive-curvature solution. Therefore, it reproduces Friedmann's Big Crunch scenario in which the universe will expand for a time, reach a maximum radius, and then contract.

In calculating the age of the universe (since it began expanding), Einstein made a numerical mistake, as recently noted by scholars Cormac O'Raifeartaigh and Brendan McCann.[8] Although he ended up with an estimate of roughly ten billion years, his figures didn't justify that number and predicted instead (if calculated correctly) an age of about two billion years, a much lower age than the age of the earth, the sun, and most of the stars we see today—which, of course, doesn't make sense. One reason for the low estimate (with the correction) is because the value of Hubble's constant was inexact owing to the limits of observation at the time. In the intervening decades, modern techniques have established its value more precisely. Yet, even if we correct Einstein's calculations, and supply a contemporary value for Hubble's constant, Einstein's 1931 model still doesn't quite get the age of the universe right because it doesn't take into account the presence of invisible substances called dark energy and dark matter, each of which drives cosmic dynamics. Also, it assumes an overall positive curvature of space, whereas current observations indicate that the universe is flat instead.

A major clue as to the source of Einstein's computational error is a blackboard housed at Oxford's History of Science Museum that preserves what Einstein wrote during a public lecture on cosmology that he delivered at Oxford University on May 19, 1931, as part of the Rhodes Lecture Series. For years, the blackboard was displayed without properly identifying on which cosmological theory the associated lecture had been based. O'Raifeartaigh determined that the equations Einstein wrote matched his first published expanding universe model. Moreover, by checking the calculations, O'Raifeartaigh found that Einstein seemed to have erred by a factor of ten in converting the Hubble constant from one unit system to another, which threw off the value of the radius and age of the universe.[9] So, take heart if you ever tabulate a bill incorrectly—even Einstein occasionally made silly mistakes in math.

An even more significant finding by O'Raifeartaigh pertained to an unpublished work by Einstein that was titled in German "On the Cosmological Problem." This manuscript is nearly identical to the April 1931 paper "On the Cosmological Problem of the General Theory of Relativity." Prior scholars likely presumed the unpublished paper was simply a draft of the published article. On the contrary, O'Raifeartaigh was astonished to discover that it presents a completely different model, one that seems to anticipate the steady-state theory of the universe. While it accounted for Hubble's discovery by modeling an expanding universe, it retained the cosmological constant and the constant overall density of matter in the universe by supposing that new particles were continuously being created. In other words, although the cosmos expanded, it would maintain the same overall appearance over time.

O'Raifeartaigh recalled his Eureka moment when he realized he had discovered a previously unidentified work by Einstein: "I leapt from my chair and, printout in hand, ran down the corridor to the office of my colleague Brendan McCann, a mathematics lecturer with a formidable command of German, exclaiming 'I think I've found something!'"[10]

In science, ideas that are absolutely original are rare. Notions are often rediscovered again and again by various thinkers, taking different shapes, until finally—thanks perhaps to a particularly cogent argument made by a persuasive spokesperson or the prospect of experimental verification—they become widely known. Therefore, perhaps it is not surprising that Einstein toyed with the steady-state concept that would later be developed independently by Hoyle, and in a different way, by Bondi and Gold around the same time.

One might wonder, then, what attracted Einstein to a universe of continuous creation, and why he abandoned the model that anticipated the steady-state without even publishing it. Given his predilection for static cosmologies, clearly a lure would have been the possibility of maintaining a steady density of matter and energy while accommodating Hubble's findings that strongly suggested cosmic expansion.

Gut feelings about what would make nature's laws seem elegant and simple are hard to dismiss. Einstein often imagined how a divine being might have crafted fundamental rules for the universe. Uniformity over space and time seemed an ideal quality. In developing general relativity, he pictured space and time on equal footing, as a united, four-dimensional space–time. Therefore, even after accepting the notion of expansion in space, he sought a way of continuing to fill up space slowly with new matter to keep it uniform over time. That would restore the idea of overall cosmic sameness in both space and time. Hoyle, Bondi, and Gold would later arrive at the same notion — on their own, also in an intuitive fashion — and similarly find it too compelling to set aside.

O'Raifeartaigh and his collaborators outlined the likely reason Einstein gave up the continuous creation idea.[11] Based on the way Einstein set up the equations for his model, there are no physical solutions other than a completely empty universe. The major problem was that he didn't include a factor accounting for how new material could be generated in space to fill the gaps left by the receding galaxies. It would take much more thought and perseverance — which Hoyle would certainly provide with his own endeavor much later — to get a steady-state model to be theoretically viable.

Perhaps Einstein's most successful venture into cosmology was yet another paper, "On the Relation Between the Expansion and the Mean Density of the Universe," coauthored with de Sitter and published in 1932. In it, they omit not only the cosmological constant but also the positive curvature (hyperspherical space) present in Einstein's earlier models. Instead, the Einstein–de Sitter universe is flat, meaning that it would expand forever. Because of its simplicity, the model generated considerable and lasting interest.

Einstein met with Lemaître several times throughout the early 1930s, both in the United States and Belgium, to discuss their mutual interests in cosmology and their various ideas. Although both were wedded to an expanding cosmology, Einstein, like Eddington, didn't like Lemaître's concept of a primeval atom. Einstein's distaste was mainly for the singularity (point of infinite density) that would start off such a universe. He didn't like indeterminate "loose ends" in science. He suggested Lemaître modify the model to avoid any initial singularity, but that turned out not to be feasible.

Nevertheless, Einstein nominated Lemaître for the Francqui Prize, the highest scientific honor in Belgium; Eddington was a member of the selection board. Awarded by the king of Belgium, Leopold III, in 1934, this prize was one of the many distinctions Lemaître received for his groundbreaking theoretical work.

BRINGING COSMOLOGY TO LIFE

In January 1933, Princeton physicist Howard P. Robertson published an influential review article, "Relativistic Cosmology," that would motivate Hoyle and many others to further the field. It summarized the rich panorama of isotropic, homogeneous models developed up to that point. As a result of analyses by Robertson and British mathematical physicist Arthur G. Walker, who had studied for a time under Eddington, that established the complete set of solutions, the most general representation of isotropic, homogeneous cosmologies has come to be known as the Friedmann-Lemaître-Robertson-Walker (FLRW) metric. It would become the geometric basis of the spatial dynamics of the Big Bang theory.

Note, though, that there are far more aspects to physical cosmology than how space expands. The universe is laced with material and full of intricate structure, from stars to galaxies and even larger arrangements such as clusters. To build that structure from scratch requires a model of how conditions in the cosmos at various stages of its development affected its content. General relativity was not enough to furnish that information and was supplemented by the fields of thermal physics, atomic physics, nuclear physics, particle physics, and related disciplines.

Eddington had started the ball rolling with his 1926 treatise, *The Internal Constitution of the Stars*. In it he suggests that the union of four hydrogen nuclei transforms them into helium nuclei and releases energy in the process (via Einstein's famous mass into energy conversion equation, $E = mc^2$). This conversion fueled the sun and other stars. His rudimentary analysis of energy-producing stellar fusion was insightful but incomplete; he proposed it well before the atomic nucleus was much understood. Not all of its constituents were even known at that point. James Chadwick would discover the neutron in 1932, which changed the whole field. It took a new generation—the likes of Gamow, Hoyle, and their associates—to probe the mysteries that united the very large with the very small: astrophysics and cosmology with nuclear and particle physics.

Attending Cambridge, starting in the autumn of 1933, offered Hoyle many opportunities to forge such connections. Although before university his main scientific interests were chemistry and astronomy, as he attended lectures by some of the greatest minds of his day, he began to gravitate toward physics and mathematics as well. In particular, Eddington's talks on general relativity were eye-opening to him.

Hoyle wasn't able to speak directly to Eddington then, but they finally interacted seven years later, when Hoyle was already in a career position at St. John's College, Cambridge. He needed a signature from Eddington for a matter pertaining to the Royal Astronomical Society, a task he thought would take less than a minute. He wasn't sure whether Eddington remembered him from the lectures or from his answers on a comprehensive exam on general relativity that Eddington had marked, but they ended up getting along well and chatted for about two hours.[12]

During their conversation, Hoyle and Eddington discussed the question of how stars are fueled in their later stages. Eddington had a well-known theory about stellar evolution, but it had a problem when applied to the sun. The temperature of the solar core was simply too low for four hydrogen atoms to simultaneously fuse into helium, as his model suggested, giving off light energy in the process. When his colleagues had pointed out that issue, he had snarkily responded that they should look for a hotter place.

Interestingly, during Hoyle and Eddington's discussion, Gamow's research came up. In a paper, Gamow had suggested hydrogen-to-lithium as a nuclear process that could transpire at relatively low temperatures. Hoyle explained to Eddington that he was dubious about Gamow's idea because there wasn't enough lithium around to suggest that the process happened very often.

The fact that Gamow came to mind during a discussion in Cambridge was an indication of how much his star had risen (pardon the pun) since his student days at the University of Leningrad. While Hubble, Lemaître, Einstein, and others were advancing the observational and mathematical descriptions of an expanded universe, Gamow had been traveling around Europe and the United States offering groundbreaking contributions to nuclear physics. In the process, he had made a name for himself at Cambridge and elsewhere. These breakthroughs would ultimately prove instrumental in advancing astrophysics and cosmology.

CHAPTER THREE

Unlocking the Nucleus

> This is the atom that Bohr built.
>
> This is the nucleus that sits in the atom that Bohr built.
>
> This is the drop that looks like the nucleus that sits in the atom
>
> that Bohr built...
>
> —RUDOLF PEIERLS, "The Atom That Bohr Built," *Journal of Jocular Physics*, 1955

A LONGSIDE THE GREAT STRIDES MADE IN COSMOLOGY DURING the 1920s and 1930s came extraordinary advances in atomic and nuclear physics brought about by the development of quantum mechanics. Although contributions to those fields were a global effort, three of the most important centers for development were Copenhagen and Göttingen, for theory, and Cambridge, specifically the Cavendish Laboratory, for experimentation. The research done in these centers would prove critical to further work in astrophysics and cosmology.

Although today it seems obvious that the sun and other stars shine because of nuclear processes and that elements are forged in their cores, such connections weren't always straightforward. Because their core temperatures were not high enough for simple models such as

Eddington's to work, scientists scratched their heads when faced with the enigma of how hydrogen nuclei overcame their electrical repulsion to fuse together and become helium nuclei. Higher up the element ladder, no one could explain how the equivalent of three helium nuclei could fuse into carbon. Thus, it would take the genius of thinkers such as Gamow (addressing the hydrogen fusion question), German-born physicist Hans Bethe (further exploring the fusion sequences in stellar processes), and later Hoyle (addressing the carbon production question) to make the intuitive leaps needed to unite astrophysics with nuclear physics and explain stellar energy and element formation. Gamow introduced Bethe, who was originally purely a nuclear physicist, to that emergent interdisciplinary field, and Hoyle followed in their footsteps. Arguably, therefore, the chain of events that led to nuclear astrophysics began with a journey by Gamow to the quantum kingdom, visiting three of its capitals in succession: Göttingen, Copenhagen, and Cambridge.

In 1928, when travel from the Soviet Union to other countries was challenging but still feasible, Gamow was offered the career-bolstering opportunity to attend summer school in Göttingen, where lectures were held during the university's break between semesters. He was thrilled to experience a foreign country, practice his basic German (which at that point was sometimes muddled with his basic English), and explore a historic city that housed a university famous for its mathematics and science. It was the city that had known such mathematical luminaries as Carl Gauss, one of the developers of non-Euclidean geometry, Hermann Minkowski, who had proposed the concept of space–time as a united, four-dimensional entity, and David Hilbert, who had played a vital role in developing the mathematical frameworks of quantum mechanics and general relativity. Indeed, the term *quantum mechanics* was coined there by the innovative physicist Max Born. Two years before Gamow's visit, Born helped unite the two versions of the theory, matrix mechanics by Werner Heisenberg and wave mechanics by Erwin Schrödinger, into a single, successful description of the workings of atoms and other systems with electrons. Gamow knew the theory well and would soon put it to excellent use in an application to nuclear physics that not even the founders of quantum physics had anticipated.

BREAKING BARRIERS

Gamow enjoyed perusing the University of Göttingen's extensive library. One day, he came across an article that would lead him to one of his greatest achievements: an explanation of how positively charged particles enter and leave a positively charged nucleus, given that like charges repel each other. Written by Ernest Rutherford, renowned experimental physicist and director of the Cavendish Laboratory, it described the scattering of fast-moving alpha particles (positively charged helium nuclei) from uranium samples. Somehow, in the process, the particles breached an energy barrier that should have been strictly forbidden for them to cross. The issue pertained to an even more fundamental question: How do alpha particles released by radioactive nuclei such as radium-226 and uranium-238 cross the energy thresholds needed for them to escape? This is a process known as alpha decay.

The situation was akin to a roller-coaster car that speeds down the first hill on the track and then encounters a second hill that is much higher. Though it builds up energy during its descent, that is not enough to propel it to even greater heights. The farthest it could climb up the second hill would be to the height of the first—and then, only if there was absolutely no friction to hold it back. Similarly, a positively charged alpha particle, when faced with the energy barrier of electric repulsion, should simply bounce away if it doesn't have enough energy of motion to cross it. But that's not what Rutherford's group at Cavendish had observed.

Rutherford, though not a theoretician, often liked to attempt his own explanations. In that way, he was very different from Hubble, who left theory to the theorists. Rutherford's idea was that the alpha particles were escorted past the barrier, in both directions, by a pair of negatively charged electrons that exactly balanced the alphas' positive charges, making the whole projectile electrically neutral. The electrons would drop back into the nucleus once their job was done, so that's why researchers didn't detect them. Therefore, the neutralized alpha particles were not hindered by any barrier of electrical repulsion.

Gamow found Rutherford's explanation far-fetched. It seemed too convoluted to suppose that a pair of electrons would serve a function for the exact fleeting interval when they were needed and then just as rapidly leave the scene as soon as they weren't. To return to the analogy, it would be as likely as a rogue geyser erupting under the roller-coaster tracks at just the right instant to propel the struggling car up the second hill and then disappearing back into the ground as soon as its job was done.

Some theorists hunt down the most complicated problems and spend months or years developing elaborate solutions. Gamow was never that kind of theorist. Rather, he loved to take a swipe at the low-hanging fruit in science—ingeniously spotting them and reaching them in clever, intuitive ways. As his career developed, he would provide the hunches, and his students or colleagues would perform the actual calculations needed to support them.

Gamow remembered examples from quantum mechanics in which electrons were able to tunnel through energy barriers in a manner that classical physics prohibited. The French physicist Louis de Broglie had shown that electrons could be described as "matter waves": globs of electric charge spread over a region of space. Schrödinger had discovered an equation matching de Broglie's idea that depicted how such "wave functions" (a term denoting how such waves are spatially distributed) behave in the presence of forces. Then, to reconcile Schrödinger's equation with a theory of transitions developed by Heisenberg, Born had reinterpreted the wave function as a "probability distribution" rather than a matter distribution. That is, it represented the *chances* of an electron being located at any point in space. In some cases, the energy barriers certain forces presented allowed the tails of wave functions to get through—meaning, the corresponding particles had a chance of crossing—even though, in classical physics, the walls would have blocked the particles. (Mathematically, the tails represent exponential decrease, like an ice cube becoming increasingly smaller in the hot sun.)

Gamow rapidly drew up a model of the energy barrier for atomic nuclei. Positively charged nuclear particles like to dwell in a tiny region (within the nucleus as a result of what is now known as the "strong interaction"), lose their affinity if separated a certain amount (because of electric repulsion), and finally regain their freedom once they are far enough apart (due to the electric force decreasing with distance). By plugging into Schrödinger's equation, Gamow was able to map out the wave functions of alpha particles in a nucleus and show that they possess "tails" that extend beyond the barrier. He was able to calculate from his model the rates of decay for alpha particles and their chances of crossing nuclear energy barriers.

In typical Gamovian style, quick and impressive, but not always comprehensive, he completed the whole problem overnight and shared the results with Eugene Wigner, a brilliant assistant of Hilbert, born in Hungary, who would later go on to win the Nobel Prize in Physics. Wigner was extremely impressed. Gamow published the results to great acclaim, only to find out that a pair of physicists working at Princeton, Ronald Gurney and Edward Condon, had solved the same problem independently. Thus, all three are credited with solving the long-standing riddle of alpha decay. On the basis of his groundbreaking research on alpha decay, the University of Leningrad awarded Gamow a PhD that year.

THE HOUSE THAT BOHR BUILT

On the way back from Göttingen to Leningrad, Gamow paid a quick visit to Copenhagen. He thought it might be his one and only opportunity to visit the mecca of quantum physics and meet its chief sage, Niels Bohr.

Niels Bohr Institute for Theoretical Physics, Copenhagen, Denmark. CREDIT: Photograph by Paul Halpern.

When he arrived at Bohr's Institute for Theoretical Physics without giving any advanced notice, he asked Bohr's secretary, Betty Schultz, if he could have an appointment. When she replied that it might take a few days, Gamow explained that he had the funds for only a single day in Copenhagen. Upon hearing that, Bohr opened up time in his schedule and met with Gamow that very afternoon.

The two physicists hit it off right away. Gamow explained his theory of alpha decay and tunneling, which piqued Bohr's interest. Much to Gamow's surprise, Bohr offered him a yearlong Carlsberg Fellowship (funded by the famous Danish brewery) right on the spot. Gamow heartily accepted. He would use the opportunity to further his research into alpha particle processes, including decay and collision.

In short time, Gamow began to feel right at home in Copenhagen. He loved the fact that "the Institute buzzed with young theoretical physicists and new ideas about atoms, atomic nuclei, and the quantum theory in general."[1]

One of Gamow's major contributions that year was the "liquid drop" model of the nucleus. He proposed it as a way of understanding how the nucleus is held together and under which circumstances certain particles or clumps of particles might break off in the process of radiation. His intuitive idea, which he suggested to Bohr, was to imagine the nucleus as a thick, incompressible fluid. Picturing the physical factors, such as surface tension, holding a drop together offered insight into how the nucleus is bound. Bohr embraced the notion and later developed it further with the American physicist John Wheeler.

As the months in Copenhagen flew by, Gamow noticed that the researchers working at the institute adored Bohr and found his quirks rather amusing, especially his frequent obliviousness. For example, Bohr loved climbing and often tried to show off his skills. One day, he told some of the researchers that he would show them how to scale the walls of a building. As everyone watched in amazement, he started climbing up the side of a commercial establishment. Soon, however, the small crowd was joined by a concerned police officer. The building Bohr happened to pick was a bank, which the police thought he was trying to rob.[2]

Gamow brought his own levity to the institute as well. Once, when Fritz Kalckar, a young physics student, was visiting, he was jolted by disturbing popping noises coming from another room in the building. Naively, the first thought in the student's head was that Bohr was conducting dangerous experiments with atoms that were creating small explosions. When he entered the room where the noise was coming from, it turned out to be Gamow and another student playing a game of Ping-Pong.[3]

With his Danish, English, and German (the three main languages used at the institute) each in the developing stages, Gamow found that one way of making people laugh was through his caricatures, a skill he had picked up, perhaps, from his Leningrad friend Irina Sokolskya. Along with entertaining, he also soon discovered that such cartoons could be informative.

Each year, the institute held a conference, which traditionally ended with a lighthearted series of funny performances by the young researchers. Gamow applied his cartooning skills to participate on several different occasions. In 1931, the tenth anniversary of Bohr's founding of the institute, he drew a series of panels detailing the history of quantum physics that featured Mickey Mouse (mimicking the original version of the character drawn and voiced by Walt Disney) in the role of Bohr. The following year, which happened to coincide with the one-hundredth anniversary of the death of the famous German writer Goethe, researchers enacted a parody of Faust, written by scientist Max Delbrück (who would later win the Nobel Prize). In it, Bohr was portrayed as the Lord, Einstein as a king infested with fleas, and Eddington as an archangel. Physicists Wolfgang Pauli, known for his acid critiques, and Paul Ehrenfest, seen as a troubled soul, were cast as Mephistopheles (the Devil) and Faust, respectively. For a copy of the script that they'd warmly present to Bohr and the institute, Delbrück invited Gamow to provide the illustrations. Gamow responded with hilarious caricatures of a diabolical Pauli, a wild-haired Einstein, a sainted Eddington, and so forth.

Gamow was always gentle in his teasing of other scientists. He wanted them to embrace the joke rather than be put off by any mockery. As Igor Gamow recalled, "Father had a tendency to say whatever came to his brain at that moment. Father was certainly sweet and never mean-spirited. Father just loved science."[4]

Not just in his public works would Gamow display his buoyant humor. His private letters were replete with puns, silly comments, and funny doodles. As physicist Freeman Dyson recalled, "Gamow liked to communicate by hand-written letter in a delightfully personal style."[5]

One of the major pastimes of institute researchers, along with table tennis and humorous annual plays, was watching movies, particularly westerns. Unlike the murkiness of quantum reality, these entertainments offered a clear-cut distinction between two sides: gallant heroes and sinister villains. After seeing such a cowboy movie, Gamow posed a riddle to Bohr: "Why," asked Gamow, "should the hero in those movies always draw his gun faster than the villain? After all, the hero is unprepared while the villain knows in advance what he is going to do. He would move more quickly."[6]

Bohr, forever the optimist, thought otherwise. The good guy would have a much calmer demeanor, he argued. Without a guilty conscience, he'd be quicker on the trigger than the troubled villain.

As scientists, they knew they needed to test their hypotheses. The next day, Bohr and Gamow formed two teams—the heroes and the villains, respectively—each armed with toy pistols. Gamow's squad went into hiding and launched a sneak attack on Bohr's posse. Each drew guns, but Bohr fired first. He was right!

The love of western movies and their cowboy heroes inspired Gamow to Americanize his Russian nickname "Geo." He insisted, from that point on, that everyone should spell it "Geo" but pronounce it "Joe." As Igor Gamow explained: "My father was tall and thin. 'Joe' was the hero. The cowboy always got his gun out faster. Father was always passionate about horses and cowboys."[7]

Gamow's identification with cowboys also reflected his unorthodox approach to science. He loved to venture like a solitary horseman out to new frontiers and make his mark. Rather than settling in that outpost, however, he'd simply move on to another—forever the lone ranger.

RENDEZVOUS WITH RUTHERFORD

Bohr had a long-standing connection with Rutherford, dating back to when he had developed his embryonic model of the atom (imagined as electrons circling a tiny nucleus) based on the scattering data collected by Rutherford's group. He encouraged Gamow to visit Rutherford during Christmas. Knowing that Rutherford could be temperamental (one of his nicknames was "Crocodile," because of the way he snapped at people), Bohr encouraged Gamow to bring graphs showing how his theoretical predictions for alpha scattering matched some of the lab

Cavendish Laboratory, Cambridge. CREDIT: Photograph by Paul Halpern.

results Rutherford's team had obtained. They anticipated that Rutherford would be pleased to see the confirmation. The strategy worked perfectly. Rutherford was impressed, and he and Gamow got on well. Gamow returned to Copenhagen elated.

Months later, when Gamow's yearlong Carlsberg Fellowship was about to expire, he and Bohr had a meeting of minds to decide what he should do next. He didn't want to return to Leningrad full-time just yet. Traveling to Cavendish and spending a year with Rutherford seemed the perfect solution. With Bohr's endorsement, Gamow received a yearlong Rockefeller Fellowship, which would fund his stay. So, upon the close of his fruitful year on Danish soil, and a summer visiting Russia (where he was lauded in the press for his nuclear discoveries), he set off for an illuminating English adventure. (Hoyle, still in high school at that time, had yet to arrive in Cambridge for his studies, so the two wouldn't cross paths.)

Upon arriving in Cambridge, Gamow bought a second-hand motorcycle for transportation and to jaunt around the countryside during his time off. When his good friend Landau visited during the summer of 1930, they traveled through England and Scotland together on the bike. For Gamow, the vehicle served as a symbol of bold independence, alongside its function as a mode of transportation. It granted him the freedom of travel that was increasingly a rare commodity in the Soviet Union.

Not everyone felt the same way. One time, J. J. Thomson, discoverer of the electron, former director of Cavendish, and mentor to Rutherford, saw Gamow riding and gasped in horror. "Get him off his motorcycle," Thomson was reported to have said. "If he gets killed, he'll set physics back twenty years."[8]

At Cambridge, Gamow met the brilliant physicist Paul Dirac, whose contributions to the mathematical formalism of quantum mechanics were definitive, and who correctly predicted the existence of antimatter. Dirac was a fellow at St. John's College, soon to become Lucasian Professor of Mathematics, the position once held by Isaac Newton and later held by Stephen Hawking. The taciturn Dirac and the jokester Gamow had a great affinity. Whereas Dirac was a man of few words who often took long pauses to gather his thoughts, Gamow was happy to fill in the gaps. And with Dirac emotionally stymied by a restrictive childhood, Gamow brought him joy by introducing him to lowbrow pastimes. For instance, he acquainted Dirac with Sherlock Holmes stories and Mickey Mouse cartoons and taught him how to ride the motorcycle.[9]

Like Bohr's institute, Cavendish similarly had a long-standing tradition of fun annual celebrations. Each year researchers' voices would harmonize in special physics carols written by Thomson. A popular number they sung was "Ions Mine," with Thomson's lyrics set to the tune of the folk song "Clementine." It began:

> *In the dusty lab'ratory,*
> *'Mid the coils and wax and twine,*

There the atoms in their glory,
Ionize and recombine.
Oh my darlings! Oh my darlings!
Oh my darling ions mine!
You are lost and gone forever
When just once you recombine![10]

As his son Igor reported, later in life Gamow would sing his own parody of the song, about a nuclear reactor that was code-named "Clementine." One of Gamow's favorite western films, *My Darling Clementine*, starring Henry Fonda, was a likely source of inspiration.[11] But conceivably he also recalled the Cavendish caroling.

Of course, Gamow went to Cavendish to do research, not for leisure. He was able to have some fun on the side mainly because he was so quick in solving problems. He preferred finding the most intuitive resolution of a scientific riddle, and then enjoy himself, rather than expend time and energy in the nitty-gritty of lengthy calculations. Magnificent hunches were his trademark.

One of his major contributions to Rutherford's research goals came to him in a snap. The Cavendish director had ambitious plans for bombarding nuclei with various types of ions, with the goal of cracking them open like walnuts hit with a hammer. He turned to Gamow for guidance. The simplest ion, that of hydrogen, is a mere proton. Rutherford asked Gamow to calculate the ideal energies with which to hurl protons into nuclei to breach their energetic barriers with just enough oomph to split them apart. Of course, Gamow agreed.

In a flash, Gamow got back to Rutherford with the requested estimate. The answer, he explained, was that protons would require one-sixteenth as much energy as alpha particles to hit the same targets. Protons have about one-quarter the mass of alpha particles and would need to travel at half speed. Plugging those figures into the standard kinetic energy formula found in introductory physics primers quickly gave him the solution.

Rutherford was astonished by Gamow's elementary solution. "Is it that simple?" he wondered. "I thought that you would have to cover sheets of paper with your damn formula."

After Gamow assured him it was that easy, Rutherford trusted the value. He brought in his two young assistants, John Cockcroft and Ernest Walton, to review the solution. They were in the midst of constructing the world's first linear accelerator, with the goal of speeding up particle projectiles using voltage boosts in order to bash them into atomic targets. Rutherford had instructed them to pack "a million volts into a soapbox." Gamow's result proved very helpful because, ultimately, they were able to achieve the desired results with a machine that put out only six hundred thousand volts.

Inspired, in part, by Gamow's calculation, the Cockcroft-Walton experiment helped revolutionize the world of physics. Though Gamow was not around to see the project to fruition, he was delighted to hear in April 1932 that the team had bombarded lithium nuclei with protons, splitting each into two helium-4 nuclei, plus extra energy. The energy produced closely matched the mass difference between the initial and final values multiplied by the speed of light squared, providing the first experimental verification of Einstein's famous formula. Moreover, the experiment offered the first proof that elements could be transmuted from one into another: a scientific take on the alchemists' dream.

Around this time, James Chadwick, also working under Rutherford, identified the neutron, another milestone. Two years later, the revolution resulting from the Cockcroft-Walton experiment would be extended even further with the first successful nuclear fusion experiments, also at Cavendish. Mark Oliphant was able to transform deuterium (an isotope of hydrogen with a proton and a neutron) into tritium (an isotope of hydrogen with a proton and two neutrons), helium-3 (an isotope of helium with two protons and a neutron), and the more common helium-4 (two protons and two neutrons).

Recalling earlier discussions with researchers Fritz Houtermans and Robert Atkinson, who were trying to firm up Eddington's idea that solar energy derived from nuclear fusion, Gamow immediately realized the relevance of such experiments to astrophysics. With the help of Gamow (and his tunneling idea), Houtermans and Atkinson developed a simple formula to describe the fusion process in stars. Gamow began to think about ways by which nuclear reactions in stars (and eventually the Big Bang itself) could produce the chemical elements. His explorations helped create the interdisciplinary field of nuclear astrophysics—tying together the very large and the very small in an unprecedented and revealing fashion.

After his fellowship at Cavendish expired, Gamow headed back to Copenhagen, sporting his motorcycle. Bohr arranged for him to stay at the institute until spring 1931. During that time, Gamow's shiny vehicle became a source of envy. Many fellow researchers, and even Bohr himself, who was very athletic, asked permission to take it for a spin around the city. Gamow happily obliged.

Through the winter and spring of 1931, Gamow continued to relish the atmosphere in Copenhagen—a delightful, festive city, even in the darkest, coldest months of the year. In a way, he had become part of Bohr's family. Bohr's children, with whom he sometimes played ball games, called him "Uncle Gamow."[12] The only low point was a skiing accident during a trip with Bohr to the Norwegian mountains that left him with a lifetime of flareups of his injured right knee. Other than that, it was a wonderful time of his life, full of theorizing and frolicking. Sadly, though, with his visa about to expire, he needed to head back to Leningrad. He bade farewell, hoping he'd return soon to an intellectual environment in which he thrived.

ESCAPE FROM RUSSIA

Although it had been less than two years since Gamow's last visit to Russia, conditions had changed dramatically. The Soviet government, its power completely concentrated in the hands of dictator Joseph Stalin, had started to crack down on independent science. Not only were science students required to learn Marxist-Leninist theory, later as scientists they were expected to incorporate such political ideas into their fundamental research. Theories were judged on the basis of whether or not they were consistent with Communist ideology.

Moreover, travel abroad was restricted. Gamow would need to obtain a new passport, which would be hard to get. Any contact with foreigners, unless perhaps they were sympathetic Communists from other countries, was discouraged. For Gamow, who relished scientific truth and international cooperation, the regime's policies offered a kind of intellectual dead end. How could he thrive under such conditions?

But hard work was a must to make ends meet. He found that he needed to carry five different academic titles, including professor at the university, and positions at various institutes in Leningrad, such as the Radium Institute, to collect enough salary to live comfortably. All he could think about was how to get out of the country and back to Copenhagen or elsewhere.

One thing going for him in the months after his return was falling in love with a remarkable woman, the strikingly graceful physicist Lyubov Vokhmintseva, later nicknamed "Rho." They were an excellent match in terms of interests, if not temperament; throughout their relationship they'd often have heated arguments. Like Geo, Rho was attuned to culture as well as science. With bohemian proclivities, she pursued poetry, photography, and ballet dancing. They met at a lively party, got married shortly thereafter, and were inseparable for many years.

But faced with an increasingly oppressive society, soon Geo and Rho were plotting their escape from Russia. They hatched a bold scheme to travel south to the Crimea, launch a small boat from a port town, cross the Black Sea (a distance of 170 miles from their planned embarkation point), and flee to Turkey. From there, they intended to claim that they were Danish citizens (he could speak enough Danish to get by) and ask to be taken to the Danish embassy in Istanbul, where he'd phone Bohr and request help.[13]

Their wild escape plan started off promising enough. Gamow managed to procure a rubber kayak and bring it to the intended place of departure. Filling it with enough food for five days, plus a couple of bottles of brandy to help pass the time, they set off from a Crimean dock as intended. On the first day, under calm skies, they made considerable headway. Unfortunately, the weather then turned stormy. Battling for their lives, they managed to make it back to shore, still in Crimea, landing about sixty miles from where they had set off. At least, though, they made it back intact—with nobody suspecting that it was anything but a pleasure outing gone awry.

The Gamows also considered fleeing from the northern border, near the city of Murmansk. Some of the Sámi, an indigenous people of that Arctic region, happily offered to guide would-be refugees across the frozen tundra to the tip of Norway and freedom. But after some research, Geo realized that many of those so-called guides actually made much of their money from Soviet border guards. They'd collect the funds from clients, betray them to the guards, and gather a reward. Geo and Rho were smart enough not to fall for the bait.

Meanwhile, Bohr started to worry about Gamow's lack of response to conference invitations and other correspondence. Gamow missed an important conference in Rome that he really wanted to attend but that the Soviets wouldn't allow for political reasons having to do with Soviet-Italian relations. Bohr correctly sensed that something was wrong. A major conference on nuclear physics was coming up in 1933, the seventh installment of the renowned Solvay series of meetings in Belgium. It would be a great loss if Gamow wasn't present. Bohr thought hard about how to get him there.

The solution was to extend an official invitation to Gamow as the Soviet representative to the Seventh Solvay Conference. Bohr arranged for the conference chairman Paul Langevin—a noted French physicist who also happened to be politically active on the left and openly sympathetic to the French Communist Party—to issue the request directly to Soviet officials in Moscow. With an official imprimatur, it was bound to be effective.

Gamow was stunned when, after all his efforts to flee illegally, the Moscow *apparatchiks* seemed to be offering him a golden opportunity to travel abroad. The major problem he faced was that the invitation to Solvay was for him alone. He absolutely couldn't leave without Rho. In past years, bringing one's wife to a conference was relatively straightforward, but the authorities had clamped down. Thus, it took him a long time, including meeting with top government officials such as Nikolai Bukharin (in charge of science policy) and Vyacheslav Molotov (second in command to Stalin), to convince the Soviets that he needed her to come along. Finally, after threatening *not* to go, and thus perhaps embarrassing the Soviet regime, he and Rho were offered new passports and permission to travel. They seized the rare opportunity. Soon they scurried toward the Finnish border and onward to Copenhagen for the conference. Gamow would never see his homeland again.

The Solvay Conference of October 1933 was noted not only for who was there but also for who wasn't. While Bohr, Heisenberg, Rutherford, Chadwick, Cockcroft, Walton, German-born nuclear physicist Rudolf Peierls, and pioneering French scientist Marie Curie were in attendance, among others, Einstein was notably absent. The Nazis had come into power in Germany earlier that year, enacting repressive policies and attempting to confiscate Einstein's possessions while he was visiting Caltech and en route back to Europe, so Einstein had decided not to return to Germany. He resided in Belgium until that became too dangerous because of Nazi threats against his life, moved briefly to England, and finally took up a position at the newly founded Institute for Advanced Study in Princeton, New Jersey. He moved there right before the Solvay Conference and would never return to Europe. Peierls, who, like Einstein, was Jewish, also needed to escape from Germany and chose England as his final destination.

After the Solvay Conference, Bohr encouraged Gamow to return to Russia, so as not to embarrass Langevin, who had gone to great lengths to issue the invitation at Bohr's request. Gamow discussed his situation with Curie, who in turn had a long talk with Langevin. The gist of what they decided was that Gamow could spend some time working in the Curie lab in Paris, which would offer him time to find a new position outside of Russia. Perhaps thinking about the fate of Europe under the threats of Stalin, Adolf Hitler, Benito Mussolini, and others, Bohr then advised Gamow to relocate to the United States.

Dutch quantum physicist Samuel Goudsmit, who had relocated to the United States several years earlier, had already invited Gamow to visit the University of Michigan in Ann Arbor for a few months. So, Gamow wrote to Goudsmit to accept the offer, believing that university would be a good launching point. In the meantime, because the visiting appointment was for summer 1934, he spent several months in Paris, Copenhagen, and Cambridge, until the time was right for him and Rho to make the transatlantic crossing.

After they arrived safely in Ann Arbor, Goudsmit offered Gamow a warm welcome. Once Gamow had settled in and gotten to know the sprawling university campus, the Dutch physicist asked him for a favor. A new graduate student in mathematics named Tompkins had just arrived. Could he show Tompkins around?[14] Gamow complied. Goudsmit subsequently informed him that Tompkins and another student would be available for assisting with any mundane tasks.[15]

Gamow enjoyed wordplay. For some reason the name "Tompkins" resonated. A few years later, in 1938, he would borrow it as the surname of a character he invented for a fun series of articles—and later books—about a befuddled bank clerk who is cast into strange predicaments that illustrate weird aspects of modern science. For his fictional Tompkins's given name, he'd choose the initials "C. H. G.," standing for three fundamental constants: the speed of light, Planck's constant (from quantum theory), and the gravitational constant, respectively. Gamow not only wrote the delightful science-laden text but also illustrated the book (by tracing from patterns).[16] The "Mr. Tompkins" series became a milestone in popular science writing.

COPENHAGEN ON THE POTOMAC

As Gamow was scouting for positions, opportunity finally knocked. While visiting Michigan, nuclear engineer Lawrence Hafstad conveyed a promising message from his colleague Merle Tuve of the Carnegie Institution for Science in Washington, DC. Tuve and Hafstad, along with another scientist, Odd Dahl, had been bombarding lithium with protons accelerated by a powerful Van de Graaff generator (a voltage-producing device) in order to look for nuclear by-products. Their experiments had commonalities with the work of Rutherford's group, especially that of Cockcroft and Walton, only they were using a different kind of particle accelerator. Tuve wanted to hire a theoretical nuclear physicist to help interpret his experiments but didn't have the funds. He had reached out to the president of George Washington University (GWU), Cloyd Heck Marvin, suggesting that a theoretical physics position be established there to support the work of both institutions and create a bridge between the two. Marvin had agreed, and the two then decided that Gamow would be a great candidate.

Gamow was excited and asked Hafstad to convey his interest. Soon a letter arrived with an invitation from Marvin to visit GWU, located in the Foggy Bottom district of Washington, DC. Gamow saw "Washington" and immediately ordered tickets to travel to Seattle, in the state of Washington. Right before departure for the Pacific Northwest, he realized his blunder and changed the destination to the nation's capital on the East Coast instead.

Judging from the way Gamow had navigated the Soviet bureaucracy arguing for permission for his wife to travel with him to Solvay, he had excellent negotiating skills. He had to put them to good use in negotiating with Marvin about the GWU position. He convinced Marvin that it really should be *two* positions: one for him and another for a second prominent theorist with whom he could collaborate. Once Marvin accepted, Gamow suggested for the second hire Hungarian physicist Edward Teller. Gamow had met Teller in Copenhagen during the spring of 1931, when the two had gone on a motorcycle adventure through Denmark, and again in the spring of 1934.

The two emigrant theorists were a good fit. Gamow loved to bounce ideas off Teller, hoping some would stick and inspire him to complete productive calculations. One of their most successful collaborations, known as the Gamow-Teller transition in beta decay, was published in 1936. Beta decay is the radioactive process by which nuclei emit beta particles (energetic electrons). Gamow and Teller's paper offered specific rules and probabilities describing one version of that process.

Rho, meanwhile, had to adjust to a new culture because she hadn't spent much time living abroad. Each day came with surprises. One of the biggest was when she found a strange package in the mail with no return address. She opened it and out popped a baby alligator that snapped at her fingers. She wondered if it could be one of her husband's pranks. As it turned out, Dirac had been vacationing in Florida and wanted to send them a present. He eventually confessed to being the anonymous sender. The Gamows had no idea what to do with the alligator, which survived only a few months.[17]

Soon Geo fell into a daily routine. Typically, he'd call Teller in the morning and share his thoughts about potential new research directions. Teller was very patient listening to Gamow's ideas. He had to be. Most of Gamow's suggestions went nowhere, as his son would describe: "Teller was very close to Father. Every morning Father would call Teller with a new idea and it was always wrong."[18]

That son, named Rustem Igor Gamow, but generally known as Igor, was born on November 4, 1935. The family bought a house in Bethesda, a suburb of Washington. One time, Einstein came for dinner, which gave Igor lifetime bragging rights that he had met the genius. However, he didn't actually remember much of the dinner because he was so young at the time, he was still sitting in a high chair.[19] The Gamows and

Tellers often got together socially. Rho would become especially close to Edward's wife, Augusta Maria, known as "Mici."

One of the rare points of tension occurred when both families took a Christmas vacation together in Miami, Florida. While soaking in the warm rays on the beach, Geo and Rho got into a heated argument. Later, over dinner, Rho blurted out to the Tellers, "Geo is really anti-Semitic, and he can't stand all the Jews in Miami."[20] Her comment embarrassed her husband considerably, especially since Teller was Jewish and had fled anti-Semitism in Europe.

As Igor noted in hindsight, the question of whether or not Geo was anti-Semitic was complicated. When talking about someone new, his father would often ask, "Is he Jewish?" On the other hand, the majority of his good friends were Jewish, from Landau to Teller, and he adored them.[21] After that tense vacation dinner, Teller decided to have a talk with Gamow. In their conversation, Gamow recalled those in the Bolshevik movement who were of Jewish heritage, such as Leon Trotsky and Grigory Zinoviev, who had created so much turmoil in his native land. "He turned out to be more anti-Communist than anti-Semitic,"[22] Teller concluded, with a sense of relief. The subject reportedly never came up again.

As a second condition for acceptance of his appointment at the university, Gamow had asked Marvin to establish a yearly physics conference at GWU, cosponsored by the Carnegie Institution. Gamow's aim was to bring the "spirit of Copenhagen" to Washington to promote fruitful discussions among international theorists. Marvin had enthusiastically agreed. Teller played a major role in planning the annual scholarly conferences; part of his job, therefore, was conference organizer.

The Washington Conference on Theoretical Physics first convened in 1935. It would assemble annually until 1947, except for a three-year break during World War II. Topics included nuclear physics, molecular physics, low-temperature physics, biophysics, and astrophysics. As intended, the meetings attracted luminaries from around the world.

A frequent presenter was Cornell nuclear physicist Hans Bethe. Born in Germany, he had left his native land because of the Nazis. The fourth Washington Conference, held in 1938 with a focus on stellar energy and nuclear processes, inspired him to switch his research pursuits from standard nuclear physics, a field to which he had contributed seminal work about deuterons (the result when protons and neutrons combine), to follow in the path of Eddington, Houtermans, Atkinson, Gamow, and others in trying to unravel the question of how stars are powered by nuclear fusion. In that domain, he would make so much progress that his 1939 series of papers, "Energy Production in Stars," has become essential reading in astrophysics.

One of Bethe's key findings, motivated in part by discussions at the 1938 conference, was that stars fused hydrogen into helium via one of two distinct processes depending on their mass and the chemical composition of their core. The first method, transpiring primarily in less massive stars, is called the proton–proton chain in which two protons combine (and one undergoes beta decay) to create a deuteron, which melds with another proton to create helium-3, which merges with yet another proton (and one undergoes beta decay). As recent studies have shown, this process provides about 99 percent of the sun's power. The second method, occurring mainly in heavier stars that have sufficient quantities of carbon-12, is known as the carbon–nitrogen–oxygen cycle. It relies on those elements to catalyze the process of transformation by gobbling up protons, the nuclei of hydrogen, and ultimately (at the end of each cycle) expelling alpha particles, the nuclei of helium-4. About 1 percent of solar energy stems from that method. One of the mysteries Bethe didn't explain was how that carbon-12 was created in the first place. In the years ahead, other physicists, including Hoyle, would finally solve that puzzle.

The most famous of the Washington Conferences, the fifth meeting, held in 1939, featured a rather dramatic announcement by Bohr. Although the main thrust of the meeting was low-temperature physics, he announced to the attendees that Otto Hahn and Fritz Strassmann, working in Nazi Germany, had split uranium nuclei in a successful fission process. Gamow had been warned of this event the day before and phoned Teller with the message, "Bohr has gone crazy. He says uranium splits."[23]

Many conference-goers immediately saw the implication that the Nazis could start a bomb program. The ensuing alarm inspired Hungarian physicist Leo Szilard to draft a letter to President Franklin Roosevelt, which Einstein and he would sign and send, warning about the dangerous potential of the Nazis developing an atomic bomb. The Manhattan Project, the American-led program to create and build the first nuclear weapons that resulted from that warning, would soon begin, with Bethe and Teller playing pivotal roles.

Although in his heart Einstein was a pacifist, the Nazis' reign of terror since their coming to power in early 1933 convinced him that they needed to be stopped at all costs. His fellow physicists, many of whom had fled that murderous regime, couldn't have agreed more. Europe was in turmoil, and the physics community, along with all concerned individuals around the world, felt a grave responsibility to act.

THE CAMBRIDGE CRASH

In October 1933, at the start of an era of great uncertainty for European physics, Hoyle became a student at Cambridge. Because of Hitler, many prominent physicists from continental Europe had emigrated. While some resettled in Britain or Ireland, such as Rudolf Peierls and Max Born (who was dismissed from his position because of anti-Semitic laws), most, such as Teller and Bethe, ended up in the United States, which effectively became the new working center of science. For instance, Hoyle arrived at Cambridge the same month Einstein settled in Princeton.

That time also marked the close of the first wave of scientific cosmology. Einstein largely stepped away from the subject and turned to other interests. Perhaps Robertson's review article, "Relativistic

Cosmology," published that year, made cosmology seem like a closed book. De Sitter died in 1934, leaving Lemaître as the main bearer of the flame for many years.

Also, the miraculous engine of quantum mechanics was starting to turn after a decade of incredible innovation. After "Thirty Years That Shook Physics," as Gamow titled a popular history of the subject, progress slowed. Most theorists had turned to nuclear physics, but development in that field was stymied by the need for increasingly powerful accelerators to generate novel results. Economic conditions of the 1930s and the start of World War II at the end of that decade ceded ground to the United States.

Still, for the first three years of his studies, Hoyle was not affected by the turmoil. Completing the Mathematics Tripos, he was awestruck by the quality of his professors. He also admired the work being done at Cavendish by Rutherford's team. He began to think about a career using the methods of theoretical physics and mathematics to interpret experimental results in labs such as Cavendish. In essence that would mean following in the footsteps of theorists such as Born, Peierls, Gamow, and many others, who transformed raw observations into predictive models. In Cambridge, mathematics also included theoretical physics. It was the path Newton, Maxwell, Eddington, Dirac, and others had taken toward brilliant careers deciphering the workings of the universe.

It was a robust, optimistic time for the boy from Bingley, now a young man. He felt excited and energized. The University Chess Club offered him a fun pastime, and he also joined a long-distance walking club. Soon he had good friends and a hardy physique as he traveled to various parts of Britain, including summer walking excursions to the Lake District, the Yorkshire Dales, and a challenging eight weeks in the mountainous Scottish Highlands. Once he had developed a taste for hiking in the mountains, he could not get enough.

Hiking, for Hoyle, was the perfect expression of his love of personal freedom, much like motorbiking was for Gamow. Many institutions were constricting and ephemeral, Hoyle felt, but nature remained open and enduring. So get away he did, as much as he could, reveling in the majestic beauty of the British landscape.

In spring 1936, Hoyle's final undergraduate term, he capped off his studies with courses in quantum field theory (the quantum explanation of particle interactions) with Peierls and Born, each of whom was teaching at Cambridge at the time. Born's class met every Monday. One Sunday, Hoyle's hiking group spontaneously decided to take a brisk, forty-mile walk from Cambridge to the countryside and back. It was a lot of effort, even for Hoyle's heightened stamina. The next day, he showed up in Born's class, wearily dragging along his traditional academic gown, his legs sore and burning. Embarrassingly, his slow stride had made him late, and he was forced to make his way at a snail's pace past the podium where Born was standing getting ready to teach—much to the amusement of his classmates.

For graduate studies, Hoyle asked Peierls to be his research supervisor. That would put him on track to be a theoretical interpreter of nuclear physics results from Cavendish. Peierls agreed, and all was arranged. Hoyle received his Bachelor of Arts and was eager to rise to new achievements.

But then on October 19, 1937, almost exactly four years after Hoyle's studies at Cambridge had begun, disaster struck. His hopes began to crash all around him, like rain pounding heavily on soaked autumn leaves. Rutherford was dead. A strangulated hernia had felled the mighty physicist.

With the ashes of the larger-than-life helmsman of Cavendish now interred at Westminster Abbey, close to Newton, the lab was rudderless. Cockcroft took over, but much of the magic was gone. With far greater funding, American research centers such as Ernest Lawrence's lab at the University of California, Berkeley, with its succession of increasingly powerful cyclotrons (circular accelerators), easily took the lead in research.

In 1935, Chadwick left for Liverpool University, a huge loss for Cambridge. The University of Edinburgh snatched Born in the autumn of 1936. And Rutherford's death brought more casualties. Numerous assistants left, seeing little future at the once glorious lab.

In a great blow to Hoyle's plans, Mark Oliphant, who had brought his prominent nuclear research to the University of Birmingham, successfully recruited Peierls to join him there (coincidentally, around the time of Rutherford's death). Hoyle was left without a supervisor and with a difficult choice to make. He considered following Peierls but ultimately decided to stay at Cambridge and find a new supervisor.

Meanwhile, after Hitler's remilitarization of the Rhineland, the German region bordering France, in March 1926, in flagrant violation of the Treaty of Versailles that had ended World War I, and his subsequent saber rattling toward the other countries bordering Germany, Hoyle developed a foreboding sense that a second world war was imminent. Perhaps his career would be cut short by armed conflict.

Nonetheless, he soldiered on, choosing as his research project a topic in quantum field theory that Peierls had suggested before he left. Mostly, he just wanted to be left alone to work independently. Informed that he needed a new supervisor, even if it was just a technical matter, he found one in Maurice Pryce, a young, recently appointed fellow. Meanwhile, he picked up vital experience at Cavendish helping out with technical questions (as had Gamow years earlier, and Peierls more recently) that he would cherish for life. In doing so, he got to know Cockcroft well.

Invited to join a prestigious research discussion group that met in the evenings, and then elected secretary of that society, Hoyle found himself in the position of recruiting speakers. Pryce suggested Dirac and an astronomer named Raymond "Ray" Lyttleton.

Nervous to contact the renowned physicist, who by then had received the Nobel Prize, Hoyle mustered the courage to call Dirac and issue the invitation.

"I will put the telephone down and think," Dirac said, "then speak again."[24]

Puzzled by Dirac's unusual phrasing, Hoyle waited and waited on a silent line. Finally, Dirac picked up the phone again and answered in the affirmative. Surely, with his long pauses, he was the John Cage of conversationalists.

Dirac's talk turned out to be an illuminating discussion around the idea that certain mathematical difficulties explaining the behavior of accelerated electrons could be resolved by including "advanced potentials": signals sent backward in time that slowed the particles down. The notion fascinated Hoyle and had a profound influence on other researchers who learned about it. Soon thereafter, John Wheeler and Richard Feynman, working together at Princeton, explored the idea, which led to a proposal called the Wheeler-Feynman absorber theory. Almost three decades later, in 1964, Hoyle, along with his student, Jayant Narlikar, would return to that concept in a proposal to modify general relativity in a way that better incorporated Mach's principle.

In the case of Lyttleton's invitation to speak to the discussion group, events would take an unanticipated turn. The astronomer told Hoyle that he was far too busy to give a seminar. However, as an afterthought, Hoyle casually asked him if he happened to need any assistance. Lyttleton responded by describing a project in which he was looking at what happens to stars when they pass through gaseous clouds in space. On the basis of one of Shapley's ideas, he had envisioned the star simply experiencing friction. But Hoyle had a profound insight that there was more to the story. In discussions with Lyttleton, he suggested the concept of "accretion," in which the star would mop up some of the amorphous material through gravitational attraction and begin to grow.[25] Hoyle's clever idea led to a joint paper with Lyttleton, more than a decade of collaboration with the astronomer, and a major shift in his focus from nuclear and quantum physics to astrophysics, the field in which he would make his most pivotal contributions.

Just when things seemed to be going well for Hoyle—desperately trying to complete his doctorate—the sonorous gong of bad luck struck again. Pryce unexpectedly left Cambridge for a position at Liverpool and so Hoyle was left without a supervisor once more.

Dirac, who liked to be left alone to think, had a well-known policy of never supervising PhD students. But Hoyle, anxious to finish his doctorate, made him an offer he couldn't refuse. He asked Dirac to sign on as supervisor, with the condition that Hoyle would essentially complete his projects on his own. Dirac agreed, and Hoyle was able to complete his degree in 1939, under Dirac's nominal supervision.

The best thing that happened to Hoyle around that time was meeting a charming young woman from Cheshire, Barbara Clark, who would prove to be the love of his life. In May of that year, an old friend from Hoyle's first undergraduate year, was visiting Cambridge. Richard Beetham, who had become a teacher, asked Hoyle to meet up with him at a popular casual eatery, the Dorothy Café, nicknamed the "Dot." When Hoyle arrived, he saw two young ladies sitting at the table with Beetham.

Beetham had brought along former pupils: the Clark sisters, Jeanne and Barbara. Jeanne was studying at Homerton College to be a teacher herself. Barbara, with whom Fred was immediately smitten, was interviewing to be a student at Girton College, Cambridge.

Fred and Barbara's courtship developed at rocket speed. In July, Fred drove up to visit her, and the two hit it off. By August, they were comfortable enough with each other to travel together on a romantic holiday in the Lake District. On December 28, they'd blissfully wed in Cheshire and honeymooned in the Yorkshire Dales, paying homage to each of their native counties.

SUPERNOVAE ON THE RADAR

By the time of his wedding, Hoyle had been appointed to the three-year academic position of fellow at St. John's College, Cambridge, joining Dirac and Cockcroft, who each supported his application, as a junior colleague. However, World War II broke out in September 1939, with

Main Gate, St. John's College, Cambridge. CREDIT: Photograph by Paul Halpern.

Britain soon enmeshed in the struggle to oust Hitler from the lands he had conquered and prevent him from invading Britain. Hoyle reluctantly put off the fellowship and joined the war effort.

Hoyle's call to arms came in the autumn of 1940, after months of anticipating a wartime assignment. As a civilian, rather than an officer or recruit, he was selected to join the radar division of the British Navy. To learn the craft, he was sent to Admiralty Signals School (ASS) in Portsmouth. Compared to the math and physics he had been studying, radar technology came easy to him. Then he was assigned to a secret lab in Nutbourne, near Chichester, where he worked on models to determine the altitude of an attacking plane.

By then, Barbara was expecting their first child. Geoffrey Hoyle was born on January 12, 1941. The three of them lived temporarily in a bare-bones cottage in the nearby village of East Ashling, where, because of a coal shortage, Barbara had to cook their meals on a camping stove.[26] When the opportunity arose, they moved to a better house in Funtingdon.

By autumn 1942, because of his talent, Hoyle was appointed director of a new clandestine unit called "Section XRC8," part of the Admiralty Signal Establishment (ASE) in Witley. (The ASE was essentially a renaming of the ASS, which had been moved to a more remote location for security purposes.) Austrian-Jewish physicist Hermann Bondi was appointed deputy director. He had emigrated to Britain and had been interned as a low-risk "enemy alien" (meaning a decent person from a hostile country) for the first two and a half years of the war before finally being released. Another Austrian-Jewish physicist, Thomas "Tommy" Gold, who had similarly been interned and released, also joined the team. Among their radar research projects, the group investigated a phenomenon called "anomalous propagation" in which water vapor affects short-wavelength components of signals.

Gold found a nearby three-bedroom farmhouse to live in, where Bondi would stay as well. Hoyle (and sometimes his family) often stayed over on weekdays; he and Barbara, with the responsibilities of taking care of baby Geoffrey, didn't want to move again.

After work, Hoyle, Bondi, and Gold would stay up late discussing astrophysics, particularly Hoyle's stellar accretion work with Lyttleton, which he somehow managed to continue. Bondi would soon join with Hoyle and Lyttleton in their research project. In just a few years, Hoyle, and separately Bondi and Gold, would introduce the idea of the steady-state universe.

In late 1944, Hoyle had his first opportunity to visit the United States. The British Admiralty Signals Establishment asked him to be its delegate to a conference on anomalous propagation at the Naval Research Laboratory in Washington, DC. He used that visit to forge critical connections with the American astronomy community, including Henry Norris Russell, Walter Adams, and Walter Baade, which would ultimately lead him to study the processes during supernova explosions that create and release the heavy chemical elements.

Geoffrey Hoyle reflected on his father's roundabout route during his first American visit:

Arriving three days before the conference, my father chose to visit Henry Norris Russell at Princeton University Observatory. From Russell he received a letter of introduction to Walter Adams at Mount Wilson Observatory. On his trip to the West Coast he stayed the weekend at the observatory, hosted by Walter Baade, who had discovered two distinct stellar populations and opened up the study of stellar and galactic evolution. Baade brought Hoyle up to date with three years of progress in American astronomy. They talked about the implosion of massive dying stars and the nuclear explosions triggered by stellar collapse.[27]

Just as his encounter with Lyttleton had introduced him to stellar astrophysics, his meeting with Baade familiarized him with a critical question in that field: What happens when massive stars die? Baade made him aware of the colossal power of stellar bursts, with temperatures in the hundreds of billions of degrees. He and another astronomer, Fritz Zwicky, had coined the term "supernova" to describe that explosive process. Hoyle would later show how such supernova explosions help explain why the solar system, and particularly Earth itself, is rich with the higher elements needed for life.

A related concept that was instrumental in shaping Hoyle's new research direction was Baade's notion of stellar populations. Baade divided stars, depending on their properties, into Populations I and II. (In the 1970s, Population III was introduced.) Population I stars, such as the sun, are generally younger and brighter. They're often found in the peripheral regions of galaxies, such as the outer spiral arms of the Milky Way, where the sun lies. Spectrography shows that they are rich in "metals." Astronomers use the term *metals* to designate elements beyond hydrogen and helium. So, for instance, carbon and oxygen are metals by that definition even though they don't seem at all metallic by chemistry's standards.

Population II stars, in contrast, tend to be older and dimmer. Typically, they dwell in the relatively crowded central regions of galaxies, such as the hub of the Milky Way, or in another formation called globular clusters. Astronomers call them "metal-poor," indicating that they are almost completely composed of hydrogen and helium. Baade theorized that when such stars reach their end stages, and run out of their primary fuel, their cores suddenly collapse. The implosion of a star's core, in turn, triggers an enormous explosion of the star's outer layer in a supernova burst.

Inspired by his fortuitous discussion with Baade, Hoyle decided to make it his mission to determine precisely how all the chemical elements were made in stars, particularly in Population II stars. He'd show how the enormous temperatures in the collapsed cores of massive stars led to the production of higher elements and how the scattering of the metals produced in collapsed stars helped generate the ingredients for the formation, over the eons, of Population I stars and their planetary systems. Hence, most elements on Earth would have originally come from one or more Population II stars that had exploded and released such material. The subject would absorb his time and energy for more than a decade after the close of World War II and result in seminal papers that changed the course of astrophysics.

But Hoyle was never one to pursue only a single topic at a time. Shortly after the war ended, a trip to the cinema with Bondi and Gold stimulated a bold excursion into cosmology, helping revive discussion in that field. In two separate papers—one by Hoyle, and the other by Bondi and Gold—they would develop the innovative, but ultimately flawed steady-state alternative to the Big Bang.

CHAPTER FOUR

Recurrence in the Dead of Night

THE THEORY OF CONTINUOUS CREATION

My recurring dream isn't just

a meaningless trick of the mind.

—WALTER CRAIG, protagonist in the 1945 film
The Dead of Night

FROM THE MID-1930S UNTIL THE MID-1940S, FEW PHYSICISTS published in the field of cosmology. The burst of energy that had centered on the idea of a dynamic universe had almost completely dampened out. Many researchers considered the work of Howard P. Robertson, at the start of that fallow decade, which had delineated all the possibilities for homogeneous, isotropic universes, the final word on the subject. Perhaps it would have been if it weren't for a splendid insight about an eternal universe that curiously emerged from a discussion about a horror film.

Hoyle, Bondi, and Gold had become close friends during the war and remained so afterward. Starting in autumn 1945, Hoyle, beginning his long-delayed fellowship at St. John's College, and Bondi, a fellow at Trinity College teaching mathematics, often got together in Cambridge, with Gold sometimes joining them. Gold had continued with naval research until being appointed to a position at Cavendish in 1947.

One evening, likely in 1946 or early 1947, the three of them decided to take a break from their heady discussions of astronomy and head over to a Cambridge cinema. Hoyle's love for the movies had persisted since childhood, when his mother's job was playing piano in the cinema to help dramatize the scenes of silent films. Deciphering the complex plots of thrillers and murder mysteries—staples of British filmmaking—was part of the attraction for him. As with his beloved game of chess, Hoyle tried to anticipate the moves. That pursuit also matched his attitude toward science—deciphering the world as a kind of self-driven detective, following leads that might have eluded others.

The three scientists settled down to watch a horror film, *The Dead of Night*. The movie has an unusual plot twist that connects its beginning and ending in a seemingly endless cycle. The machinations of the convoluted tale of terror made an indelible imprint on their imaginations.

The story begins with an architect, Walter Craig, arriving at a country house to which he has been invited by its owner, Eliot Foley, as a guest at a gathering but also to offer advice on design. Although ostensibly it is a gathering of strangers Craig had never met before in his life, he reveals that they all seem familiar because he has dreamed repeatedly and vividly of meeting them in that very setting. After he tells the dubious fellow guests of his foreboding that something bad will happen, they share their own nightmarish visions in a series of terrifying sequences that make up much of the film. The scariest (and most famous) of these is a ventriloquist tormented by his evil dummy, Hugo. As the horrific tales reach a crescendo, Craig—as if possessed by a demon—suddenly becomes overwhelmed by an irresistible urge to commit murder. In a wholly unexpected plot twist, he gruesomely kills one of the guests (for no apparent reason), is arrested by a police officer, and ends up in a prison cell, with a grinning Hugo mocking him and beginning to strangle him.

Suddenly Craig wakes up. It was all a nightmare. He is terrified but doesn't remember precisely why. The phone rings, and it is Foley, who invites him to visit his home (which Craig thinks would be for the first time). At that point, for some reason, Craig momentarily doesn't realize that the country house is connected with his bad dream. With a rural excursion seeming like a relaxing escape, Craig gets in his car, drives over there, and all of the events recur—implying that they'll continue to do so forever.

Hoyle, Bondi, and Gold left the cinema and returned to Bondi's rooms at Trinity. Bondi offered the others some rum he had procured from a relative. Upon discussing the film while downing their drinks, Gold suddenly remarked, "What if the Universe is like that?"[1]

Gold's comment inspired lengthy discussions about imagining an expanding universe that appears roughly the same over time, with no beginning and no ending. Like someone entering a movie theater in which *The Dead of Night* was looped over and over again to begin watching it at any random part, those viewing a "film" of the universe could start at any moment and see a similar picture. The main issue that emerged was how to fill in the gaps in the forever-growing empty spaces between expanding galaxies. The three began to think about how exactly new material could trickle into the universe and create new formations in the voids left behind by the receding galaxies. Amazingly, in a quintessential example of how gut feelings can drive science for intuitive thinkers such as Hoyle, a cinematic vision of terror led to a whole new model of the cosmos.

THE CEASELESS HATCHING OF NEWBORN PARTICLES

For all three physicists, each with multiple university responsibilities and various research projects to complete, the line of inquiry provoked by Gold's remark remained on the backburner until early 1948. In

Hoyle's case, not only had he started a new position, he also had a young family, now with two children. His daughter, Elizabeth Jeanne Hoyle (now Elizabeth Jeanne Butler), was born in December 1944.

Inspired, in part, by the discussion he had with Baade, Hoyle's primary research focus shifted to the question of how chemical elements are produced in stellar cores. In a 1946 paper, "The Synthesis of the Elements from Hydrogen," he sketched the rudiments of a notion he would later develop more fully: after stars much heavier than the sun exhaust their primary source of fuel, hydrogen, their cores shrink catastrophically and generate temperatures high enough for further nuclear fusion processes. Eventually, these heavier elements are released into space.

In a premonition of their coming cosmological disputes, Hoyle contrasted his own views with those expressed in a 1941 paper by Gamow and Brazilian physicist Mário Schoenberg (also spelled *Schenberg*) titled "The Neutrino Theory of Stellar Collapse." Gamow and Schoenberg had argued that neutrinos (lightweight, electrically neutral particles) carried off the bulk of stellar energy when a star's core catastrophically contracts. In an example of Gamovian humor, they nicknamed the neutrino release the "Urca process," after the famous Urca Casino in Rio de Janeiro, because the collapsing cores lost energy as quickly as gamblers at Urca parted with their money. In his paper, Hoyle rebutted that the nuclear reactions he was proposing would be a far more efficient way of releasing vast storehouses of power during such a collapse.

Hoyle gave a talk on the subject, entitled "On the Formation of Heavy Elements in Stars," at a conference in Birmingham on December 20, 1946. The thrust of his talk was that all the elements are built up from hydrogen, with the metals, the elements heavier than helium, created during the late stages of massive stars. His former supervisor Rudolf Peierls, who was in the audience, asked the natural question, "Where did the hydrogen come from?"

Challenged by Peierls to answer that conundrum, Hoyle began to think about ways elementary particles might emerge in the universe from pure emptiness. He saw little merit in Lemaître's primeval atom idea that everything was created at once. Hoyle considered such a massive violation of the law of conservation of energy and mass completely unphysical. Scientific rules shouldn't simply be tossed aside for convenience, he argued. Therefore, particles—perhaps neutrons, owing to their electrical neutrality—must somehow trickle into the spatial vacuum very slowly. That idea of "continuous creation" seemed to mesh well with possible solutions to Gold's question, he came to realize.

Even in the midst of challenging research questions, Hoyle still spent plenty of time with his family. By 1947, they were living in a house outside of Cambridge. Fred encouraged his son Geoff's scientific interests, including showing him his own childhood telescope and demonstrating how it worked.

"I remember my father setting up his telescope, which he bought in 1925, in my bedroom to watch the moon rise over the trees in the garden," Geoffrey recalled. "As I grew up I realized just how enormous the subject of astronomy was and how very little was understood and very little of the science proven. I also realized that my father's understanding of science was enormous and that, if I entered that profession, I would always be compared with my father, so chose to follow other avenues as a career."[2]

Although Fred emphasized science with his son when he was little, he apparently didn't do so with his daughter until she was older and expressed interest. Later in life, perhaps because of changing times, he was reportedly more egalitarian in how he interacted with his granddaughters. Elizabeth recalled: "Dads were not that involved with their children as now. That said he was a wonderful grandfather to my two daughters, and always very interested to hear their views."[3]

Elizabeth also commented on her father's working style and propensity to multitask: "The trick my father used to keep in touch with his family while he worked was to parallel process! He would sit working with us all around him but still be able to come in with comments or answering questions. He processed things in his head, and I'm hardly aware of a single day even when we were all on holiday when he didn't pick up his pad and pencil and work."[4]

When it came time, in early 1948, for Hoyle, Bondi, and Gold to write down their thoughts about an expanding universe that continuously created new matter, Bondi and Gold realized that they had a fundamental difference with Hoyle. Hoyle's idea for a steady-state model of the universe—the term the three of them coined—was to modify Einstein's equation of general relativity by adding a term called the "creation field," symbolized by the letter C. *Field* in this context means, in essence, a distribution of energy throughout space, with each point having a different value. Following the laws of quantum field theory, on a microscopic level fields randomly fluctuate, like the chance appearance of tiny ripples and bubbles on the surface of an otherwise still pond. Large fluctuations of the creation field, in Hoyle's model, led to excess energy, which, under the right circumstances could transform into new particles (following Einstein's equivalence of energy and mass).

Over time, the newly created particles would attract each other gravitationally, building up larger and larger clusters. Once a large enough cloud was created, astrophysics tells us that it would engender celestial bodies. If conditions were right, these would begin the process of hydrogen fusion, ignite stellar furnaces, and form shining stars. Ultimately, these stars might group together into galaxies. Thus, new galaxies would emerge to fill in the gaps left behind as older galaxies recede, lending the universe approximately the same profile over the ages. Contrast that idea with what came to be known as the Big Bang—for which nothing fills the gaps between galaxies, and the universe simply ages.

A simple analogy might illustrate the difference between Big Bang and steady-state. Imagine a nomad community residing in mobile dwellings, such as tents or cabins, and living off the land through harvesting crops. Suppose that every time they have gathered the harvest each autumn, they uproot themselves and move their dwellings far from other homes. If each household relocates in such a manner, the community would become more and more spread out each year. Larger and larger gaps would grow between the dwellings, representing abandoned land. That would be the Big Bang situation of greater and greater spacing. If though, in contrast, each community seeded the land before moving, preparing it for new harvests and new settlements in the future, and if new families moved into the spaces vacated by the old ones, the community would appear roughly the same over time. That would represent the steady-state scenario.

Bondi and Gold objected to the creation field concept for several reasons. First, they thought that the creation field would trap and agitate newly created particles, which would act like frantic flies trying to free themselves from glue. Such energetic particles would heat up to temperatures well beyond anything that has been detected in deep space. Hoyle reviewed his calculations with them and assured them that the creation field would divert particles, not trap them. Therefore, the temperatures of the newly hatched particles would not be too high. Bondi and Gold didn't buy into that explanation because it meant that the existence of the creation field contradicted some of the key predictions of general relativity. For example, globs of energy or matter should cause attraction, not repulsion.

Like Einstein in his static cosmology of 1917, Bondi and Gold hoped that a steady-state model would help fulfill Mach's principle: the notion that the combined influence of remote astronomical objects created inertia. Ultimately, all physical phenomena, they thought, should have tangible explanations. They thereby viewed Hoyle's creation field notion, which introduced an added repulsive force that no one had ever detected, as a step in the wrong direction.

DUELING STEADY-STATE MODELS

After trying in vain to get Bondi and Gold to join him in a collaborative work, Hoyle decided to go it alone. Crediting Gold for the original idea and Bondi for useful insights, he wrote "A New Model for the

Expanding Universe" and submitted it to the *Proceedings of the Physical Society* to be considered for publication.

Much to his dismay, Hoyle's manuscript was returned, along with a rather strange rejection letter. It read, in part: "Your paper has had very serious consideration by the Papers Committee of the Society who have now regretfully decided that the Proceedings is not the most suitable medium of publication, especially in view of the acute shortage of paper, which is forcing us to reject papers we would otherwise be glad to publish."[5]

The rejection letter concluded by suggesting that Hoyle submit the paper to the Royal Astronomical Society instead, implying that its topic wasn't really physics.

Meanwhile, Bondi and Gold decided to submit their own rendition of the concept, titled "The Steady-State Theory of the Expanding Universe." With an eye toward the possibility that Hoyle's paper might end up being published after theirs, they referenced his work and the discussions they'd had. Indeed, Bondi and Gold's paper would end up being published a month before Hoyle's, in the same journal, *Monthly Notices of the Royal Astronomical Society*. (Hoyle had ultimately and unhappily taken the Physical Society editor's advice to submit it to an astronomy journal instead.)

The only substantive difference between the two papers is that Hoyle modifies general relativity with the creation field term and Bondi and Gold hark back to de Sitter's empty model with a cosmological constant. In their case, the universe is not quite empty. They purport that the seeding of new matter in the universe at a rate of approximately 4×10^{-45} (a decimal point followed by 44 zeroes and the number 4) pounds per square inch per second would be so slow that it would be undetectable. Yet, as in Hoyle's model, over time, through gravity, it would accumulate into larger and larger clumps. Eventually, it would form bodies of hydrogen massive enough to ignite in nuclear processes to become shining stars. Those would form new galaxies, filling in the spaces left by the expansion of the universe, keeping its profile roughly constant over time.

Stylistically, however, the two papers are vastly dissimilar. Hoyle's exposition gets right to the point. Most of it is a systematic mathematical approach to the question of how particles might continuously emerge in a constantly expanding universe that goes on forever.

Bondi and Gold's report, in contrast, reads more like a philosophical treatise until its last few pages. Much of it is devoted to laying out the case for a brand-new law of nature that they dub the "perfect cosmological principle," which they based on a far more widely accepted idea called the "cosmological principle."

The standard cosmological principle, sometimes called the "Copernican principle," is what allows astronomers to suppose that the universe is homogeneous. It extrapolates the notion advanced by Copernicus and others that Earth is not the center of the solar system to further suppose that the solar system is not the center of the Milky Way, which in turn is not the center of the universe. Thus, our position in the scheme of things is resolutely average. Consequently, we can expect any part of the universe to look, on average, similar to our part—like imagining the scope and appearance of a vast housing development by looking out at the neighbors from inside one of the houses.

Bondi and Gold tied the idea of homogeneity and uniformity with two speculative ideas. One was Mach's principle. If inertia is the same everywhere, they argued, and it depends on the combined influences of massive objects, the distribution of the masses must be roughly similar through space.

The second idea was the "large numbers hypothesis" proposed by Dirac in a 1937 paper. His goal was to help justify an early suggestion by Eddington that the fundamental constants of nature (speed of light, Planck's constant, the charge and mass of the electron, etc.) are closely connected to each other through certain combinations that produce exact integers. Dirac noted that the ratio of the strength of the electric force to the gravitation force between an electron and proton in a hydrogen atom is about 10^{39}, and that the ratio of the mass of the universe (as supposed at the time) to the mass of the proton is about 10^{78}, which is the square of the first large number. Such a connection between two astronomically large values, Dirac thought, could not be a coincidence. Rather, he supposed, they must have a fundamental origin—which he equated with the age of the universe expressed in atomic units (minuscule intervals of time based on Planck's constant and used to describe atomic processes).

There are two cosmological ways to interpret the large numbers hypothesis. One way, embraced by Dirac himself, is that the universe is dynamic. Therefore, because the age of the universe is tied to the ratio of electricity and gravitation, a younger universe must have had a smaller ratio, meaning stronger gravitation. That is, gravitation is weakening over the eons.

Bondi and Gold's alternative way of dealing with that hypothesis is to assume that the values of the fundamental constants must be frozen and that the universe is ultimately timeless. That's where the perfect cosmological principle comes in. It extends the cosmological principle to time as well as space: not only do we inhabit an average location in space, we must similarly occupy an average moment in time. In a universal version of the movie *Groundhog Day*, every era must be the same, more or less, as every other era. As a consequence, the physical parameters producing the "large numbers" would be eternal as well.

The perfect cosmological principle similarly locks in the delicate balance of Mach's principle (presuming it were valid) for all times and all places. Inertia, the researchers argued, should be timeless. If this era's inertia derives from the combined massive tugs of distant objects, then, by golly, it should be that way always.

The appearance of Bondi and Gold's paper before Hoyle's paper served to confuse some readers. For example, in 1952, Sir Harold Spencer Jones, the Astronomer Royal and an important advocate of steady-state, wrote to Hoyle expressing the mistaken notion that Bondi and Gold were the true developers of the concept and that Hoyle's paper was an elaboration.[6] Hoyle, who worried that he'd forever be seen as an interpreter and popularizer of steady-state rather than as a co-inventor, had to explain the actual history of the concept. It would not be the last time that Hoyle's contributions were inaccurately discounted by his fellow scientists, even if, in Sir Harold's case, it was clearly a harmless misunderstanding. On the other hand, growing interest in science media would allow Hoyle to make his case directly to the public in an unprecedented manner.

MR. TOMPKINS AND THE WONDERS OF TV

In the late 1940s science expositors who offered their insights in books, articles, radio, and television were very much in demand around the world. A trend that had started in the 1920s, when newspaper mogul E. W. Scripps founded *Science Service* to supply scientific content to the press and the *New York Times* hired its first designated science reporter, took off after the dropping of atomic bombs on Japan in August 1945 at the close of World War II. The moral concerns related to the development of nuclear weapons pointed to a growing need for a scientifically educated public.

By then, Gamow had become one of the leading science popularizers in the world, and Hoyle was soon to follow. The major driver of Gamow's fame was the popularity of his Mr. Tompkins character (named, as mentioned, after a student he had met in Ann Arbor), which led him to new publishing opportunities.

Persistence reaps rewards, as we've seen for both our protagonists. Gamow's first Tompkins story, "Toy Universe," about the character visiting (in his dreams) a land in which the speed of light is only 10 miles per hour, and noticing all the strange effects of relativity in real-life situations, was rejected by several magazines, including *Harper's*, before it found the right home in the December 1938 issue of *Discovery* magazine. Gamow had learned of that opportunity from British physicist Charles Galton Darwin (the grandson of the famous naturalist, Charles

Darwin), who suggested that he send the story to C. P. Snow, *Discovery*'s editor. Snow accepted it, and Gamow's second career as a science popularizer was born.

Throughout early 1939, Snow commissioned Gamow to submit a new Tompkins story—each a different "dream" elucidating a certain aspect of science—for each monthly issue of *Discovery*. Gamow was delighted by the series' reception. Proudly, he sent personally autographed copies with various inscriptions, often humorous, to President Marvin at GWU. For example, on the cover of the February 1939 issue, featuring Mr. Tompkins's third dream, Gamow wrote, "To President Marvin. Easy reading about high-brow matters. G. Gamow."

Not to miss a marketing opportunity, by the end of 1939, Cambridge University Press collected the various Tompkins stories (thus far) and issued them in book form. *Mr. Tompkins in Wonderland*, a compendium of curious dreams about strange situations the title character found himself in, offered a delightful introduction to science. Soon Gamow landed a contract with Viking Press to write a popular introduction to stellar astrophysics, *The Birth and Death of the Sun*, which was published in 1940. *Mr. Tompkins Explores the Atom* followed in 1944.

In 1947, Gamow released perhaps his most ambitious popular science book of all, *One, Two, Three . . . Infinity*, published by Viking. The title reflected his misconception that a certain African people (then known as the "Hottentots") only had words for the first three numbers, and then referred to higher counts as "many." Despite that misunderstanding, the title offers a catchy way to convey the fact that not all counting schemes are identical. In delightful fashion, the text explores strange notions in mathematics, physics, and biology (for example, the notion that left-handed gloves can be "flipped" in the fourth dimension to create right-handed gloves) accompanied by Gamow's clever illustrations. He warmly dedicated the book: "To my son Igor, who wanted to be a cowboy."

One, Two, Three . . . Infinity was lauded for its clarity and originality. Einstein, who reportedly kept a copy of one of the Mr. Tompkins books on his shelf, praised the work for being "witty and stimulating."[7] It demonstrated Gamow's uncanny talent for mastering and explaining a vast range of scientific fields in a manner both fun and educational.

Gamow's popular science contributions didn't stop there. He would contribute numerous articles to *Scientific American* and write many more books. On television, he appeared as a guest on the *Johns Hopkins Science Review*, a program broadcast in Baltimore starting in March 1948, and spoke about cosmology (particularly about the Big Bang theory). Critics consider that series a milestone in the development of television as an educational medium.

The late 1940s was a time when television started to transition from an experimental medium to a mass medium, capable of offering entertainment and enlightenment on an unprecedented scale. In the former category was the *Texaco Star Theater*, aired by the National Broadcasting Company (NBC) beginning in June 1948. The star of the show was Vaudeville-trained comedian Milton Berle, known affectionately to viewers as "Uncle Miltie." Reportedly, his overwhelming popularity spurred many people in the United States to buy their first television sets.

Once they bought those sets, however, many viewers might have decided to switch the channel for a change of pace. In 1950, the *Johns Hopkins Science Review* was picked up by the DuMont Television Network and went national, giving viewers an educational alternative. Instead of watching "Uncle Miltie" clown around, they could listen to real scientists, such as Gamow, explain the scientific breakthroughs of the time. Given the rise of the atomic age, there was certainly a real hunger for scientific education.

It was an extraordinary experience for those at home to encounter their first taste of televised science. Less science-minded viewers could view demonstrations for the first time in their living rooms. Some of the scientists on the show received fan letters from delighted viewers. Thus, in his role as science explainer, Gamow was a broadcast pioneer, paving the way for future popular shows such as *Cosmos*.

CHRISTENING THE BIG BANG

Hoyle's foray into science popularization began about a decade after Gamow's and lasted for many years. Initially, the medium he used was radio rather than TV. Hoyle soon became a familiar voice on the British airwaves.

In the UK during the late 1940s and early 1950s, television and radio were organized differently from how they were structured in the United States. Rather than private stations, the government-run British Broadcasting Corporation, popularly known as the BBC and nicknamed the "Beeb," controlled all broadcasts. From the mid-1950s onward, private television networks, starting with ITV, were allowed to operate and create alternative programming.

One of the longest-running BBC radio shows, called *Desert Island Discs*, began in 1942 and has lasted for decades. The premise of the series is that guests are asked which eight records, one book, and one luxury item they'd bring if they were stranded on an isolated island. It is a measure of Hoyle's skills in addressing popular audiences that he was one of the few figures to be invited on the program twice, the first time in 1954 and the second in 1984. His musical choices included pieces by Mozart, Bach, Beethoven, and Schubert, a reflection of some of the classical music he was likely exposed to by his mother as a boy. The book he chose the second time (he didn't pick one the first time) was *Handbook of Physics*. His luxury items were a photograph of people at a race meet (the first time) and a portable telescope (the second time), reflecting his interests in athletic competitions (such as long-distance walking) and observational astronomy.

On March 28, 1949, Hoyle was invited to talk about cosmology on a popular radio show, the BBC Third Programme. In comparing the steady-state model of the expanding universe with other models in which the universe emerged from an ultradense point billions of years in the past, he would coin "one of the most successful scientific eulogisms ever."[8] His introduction of the term "Big Bang" disseminated when his comments were published less than two weeks later in *The Listener*, the well-read BBC programming guide. In the decades that followed, even after the steady-state idea retreated from public discourse, *Big Bang* has remained the most popular epithet for the notion introduced by Lemaître as the primeval atom and further developed by Gamow and his collaborators Ralph Alpher and Robert Herman.

As science historian Helge Kragh points out, Hoyle was not directly addressing Gamow and disparaging his newly published ideas even as he criticized what he called the Big Bang hypothesis.[9] Rather, Hoyle was speaking to those who accepted at face value Lemaître's notion of a finite cosmic age, as explored in the seminal works of Robertson, Walker, and others. The idea of an expanding universe, as most physicists had come to believe, implied that the universe was far smaller in the past. Hoyle wanted to show how that presumption was unnecessary. Regarding Gamow, Hoyle had the utmost respect for his contributions to nuclear astrophysics. Not until the mid-1950s, when both Hoyle's and Gamow's concepts of the universe were featured in the press, would their disagreements be aired.

Hoyle's son, Geoffrey, remarked: "I never heard my father say a bad word about Gamow or Lemaître. From my understanding, he felt that the Big Bang didn't solve the science of cosmology, so he continued throughout his life to revisit his own theories in the light of new information."[10]

Along with a distaste for the supreme violation of conservation laws, Hoyle had a more pragmatic argument to make against the Big Bang. The finite age of the universe, calculated according to the constant derived by Hubble from his cepheid data, appeared to be around two billion years (today we know it is really 13.8 billion years), which made the universe much younger than Earth, the sun, and most of the stars. How could a "parent" be younger than the "children" it engenders? The steady-state model, in comparison, avoided the issue of the universe's age altogether by making the cosmos infinite in duration. For that reason, on the radio show, Hoyle alleged that the Big Bang was in conflict with observation, whereas the steady-state had no such challenge.

As Alpher would later note: "In the early days there was a severe criticism of the age of the universe from the evolutionary model. It was much too short. The Hubble constant was way out of whack then, at least if you believe the evolutionary model. In fact, that's what led to the steady-state universe being proposed."[11]

When Hoyle claimed on the radio that observations seemed to contradict the Big Bang hypothesis, he didn't mention that the steady-state model was already facing its own potential challenge. In a 1948 article, astronomers Joel Stebbins and Albert Whitford announced evidence showing more distant elliptical galaxies (one of the three main classes of galactic types) have a different spectral distribution (arrangement of colors) than that of closer ones, making them even redder than what their Doppler redshifts predicted. Stebbins and Whitford interpreted those results as evidence that remoter elliptical galaxies—because the lightspeed lag meant they were farther back in time—had a different composition than closer ones. Generalized to all the elliptical galaxies in the universe, that difference implied that they had evolved over the eons—meaning that the cosmos had looked very different in the past, in stark contradiction to steady-state predictions of overall sameness.

To combat the challenge of the so-called Stebbins-Whitford effect, Hoyle, Bondi, and Gold soon found an ally in Dennis Sciama, a young fellow at Trinity College, Cambridge. Sciama, who had received his PhD under Dirac's supervision, was a firm believer that Mach's principle needed to be incorporated into general relativity. He saw the steady-state theory—with its notion of cosmic timelessness—as an important step toward that goal. In 1954, he joined with Bondi and Gold to write a key paper criticizing Stebbins and Whitford's conclusions that suggested the Stebbins-Whitford effect was cosmological. Soon after, Whitford retracted his original findings and announced that his continued investigation of the data revealed that the excess reddening had been an artifact of the spectral analysis methods he and Stebbins had applied rather than an indication of ellipticals changing over time. Steady-state was spared—at least for the time being.

Throughout the 1950s, Hoyle, Bondi, and Gold continued to promote the steady-state notion in lectures and in the media. Although its popularity in the British scientific community began to grow, it was met largely with skepticism elsewhere. Gamow reported: "According to Edward Teller it is not surprising that the steady-state theory is so popular in England, not only because it was proposed by its three (native-born and imported) sons H. Bondi, T. Gold, and F. Hoyle, but also because it has ever been the policy of Great Britain to maintain the status quo in Europe."[12]

Some of the critical voices beyond the UK included Einstein (despite having explored a steady-state-type model much earlier) and Nobel laureate Wolfgang Pauli. Einstein conveyed his feelings in a 1952 letter to the physicist Jean-Jacques Fehr: "The cosmological speculations of Mr. Hoyle, which presume the creation of atoms from space, are in my view much too poorly grounded to be taken seriously."[13]

Pauli, known to be extremely critical about other scientists' theories yet surprisingly open to the collective unconsciousness notion of Swiss psychoanalyst Carl Jung, happened to attend a talk Hoyle gave in Zürich. He made his views about steady-state known in a 1951 letter to Jung's assistant, Swiss psychoanalyst Aniela Jaffé: "I know Hoyle quite well and attended his lecture in Zürich. His mixture of fantasy and science I find in poor taste. . . . His 'Background Matter' and his continuous creation of matter out of nothing strike me as sheer nonsense. I see no reason to doubt the conservation of physical energy. It is clear to me that this kind of cosmogony is not physics but a projection of the unconscious."[14]

Hoyle and Pauli would meet again seven years later at the 1958 Solvay Conference in Brussels. With characteristic bluntness, Pauli informed him, "I have just read your novel *The Black Cloud*. I thought it much better than your astronomical work."[15]

EXPLAINING THE UNIVERSE

Hoyle continued to build a reputation as a science popularizer through numerous appearances on BBC radio, expanding beyond astrophysics and cosmology to related topics such as astrobiology. In July 1949, he conducted an on-air debate with geneticist Cyril Darlington on the subject "Is There Life Elsewhere in the Universe?" Hoyle drew upon the steady-state idea that the universe is infinitely old and endlessly vast to suggest that life could well have emerged on other planets circling distant stars at many other times in its history.

He sweetened his argument that life is commonplace with a dose of wit, remarking, "Even now, in some distant spot, another fellow called Hoyle is broadcasting to an audience like you about an identical subject."[16]

Unlike Gamow, whose humor was flung loudly and spontaneously, Hoyle's was drier and often seemed more scripted. In fact, Hoyle maintained a "joke file," full of humorous remarks that he might use to enliven lectures. In personal encounters, his humor could be rather dark (which he tried, sometimes unsuccessfully, to tone down during lectures, lest he say something offensive).

Hoyle's son, Geoffrey, characterized his style: "He had a very Yorkshire, down-to-earth sense of humour which shows up in many of his quotes. He would very often taunt academics at high table. I remember him once asking the Oxford Professor of English whether Shakespeare wrote his own work and addressing Copeland with the question, 'Was Mozart murdered?'"[17]

Astrophysicist Jayant Narlikar, who studied under Hoyle at Cambridge and later collaborated with him, recalled one noteworthy example of Hoyle's keen wit: "He was examining the answer scripts for the Cambridge Mathematical Tripos examination. The question paper had warned that complete answers must be given and fragments will not carry many marks. Despite that one student had done several incomplete questions thus collecting several sets of low marks. While I was watching Fred at work marking this script, I drew his attention to the student's name. Hoyle burst out laughing when he saw it. The name was 'Marks.'"[18]

Hoyle put his ability to charm and amaze audiences to the test when in early 1950 he wrote and delivered a series of five weekly radio broadcasts on the BBC. The shows, heavily promoted, were aired on Saturday evenings, guaranteeing them a wide listenership. Hoyle's simple, direct way of explaining tricky concepts, spoken clearly in his unpretentious, small-town West Yorkshire accent (uncommon for BBC broadcasts, typically delivered in posher received-pronunciation English), lent him a pleasant, folksy style. When transcripts of the programs appeared in *The Listener*, they were popular items. By the end of the year, Blackwell, an Oxford-based publisher, acquired the rights and reissued Hoyle's reports in the form of a book, *The Nature of the Universe*.

Hoyle's popular account of astrophysics and cosmology flew off the shelves. Readers seemed to appreciate its straightforward explanations of wide-reaching notions. It also helped bring the steady-state model into greater prominence. Yet not all reactions to the book were positive. Hoyle, in his blunt style, had made the mistake of appearing extremely judgmental about organized religion in its pages. Many devout readers were taken aback by lines such as "it seems to me that religion is but a desperate attempt to escape from the truly dreadful situation in which we find ourselves."[19]

Hoyle's book soon came up against a competitor on bookshops' popular science shelves. Gamow's own splendid take on cosmology, *The Creation of the Universe*, appeared in 1952 and also sold very well. Its message was starkly different, painting a picture of a cosmos that had emerged from an ultradense state at a finite time in the past. (Gamow never used the term *Big Bang*, which he thought was misleading because there was no "bang" in his approach. However, he did talk about the "Big Squeeze," a hypothetical contraction that produced the ylem.) It only briefly mentioned, and dismissed, the steady-state alternative. Both books were credible options as the perfect gift for science fans; the competition for adherents to the different cosmologies had truly begun.

Hoyle's growing popularity was apparent during a trip he made, starting in December 1952, to the United States, mainly to serve as visiting pofessor of astrophysics at Caltech for the upcoming spring semester but also to visit Princeton for two additional months.[20] The invitation had been issued by Walter Baade, who had sparked Hoyle's imagination eight years earlier with his supernova work. They had gotten to know each other even better at an International Astronomical Union conference held in September that year.

Before Hoyle began his work at Caltech, he was invited to be the keynote evening speaker at the annual western division meeting of the American Physical Society, held that year in Pasadena (where Caltech is located) and Inyokern (at the Naval Ordnance Test Station, to attract its scientific personnel), both in California. Hoyle's talk was a bit of an experiment, because it was the first, for meetings of that society, to be aimed squarely at the general public. It proved to be a smashing success. Hoyle's topic, "The Expanding Universe," attracted so much interest that the organizers needed to move it to a much larger venue: a lecture hall at Pasadena Junior College. Approximately fifteen hundred people packed into the auditorium, which was astonishing for a professional meeting. Undoubtedly, Hoyle's fame had crossed the Atlantic.

After the lecture, Hoyle was invited to a party at the home of William "Willy" Fowler, who conducted research at Caltech's Kellogg Laboratory.[21] The encounter would prove to be an auspicious meeting of minds. Hoyle and Fowler would become close friends and collaborators in the decades ahead.

However, at any of Hoyle's talks on cosmology, of the astrophysicists and astronomers in attendance, it is doubtful many would have endorsed his vision. Pasadena, the home of Hubble and his young disciple Allan Sandage (who was completing his PhD at Caltech under Baade, while observing at Mount Wilson), was decidedly Big Bang territory (although that term hadn't been used by that group). Even though Hubble himself

hadn't endorsed the notion of a finite-age expanding universe, his colleagues, for the most part (Swiss astronomer Fritz Zwicky being the rare
exception), accepted that view as the simplest explanation.

That said, of the cosmologists focused on a finite-age universe, not
everyone was following the research Gamow, Alpher, and Herman had
been conducting at GWU. And almost no one in that community seem
to have noticed a critical prediction made by Alpher and Herman in
1949 that radiation from the Big Bang, cooled down because of cosmic
expansion over the intervening billions of years to a frigid 5 degrees
above absolute zero, permeated the universe. Its detection in the mid-
1960s would prove to be the smoking gun for the Big Bang.

CHAPTER FIVE

Alpha to Omega

A FIERY BEGINNING

> A giant imp, jumping from atoms, to genes, to space travel, he was simultaneously there to admire when clever and to comfort when his life was going backwards. Perhaps wisely, Geo never counted on finishing the big chases he started. So he always sought fun on the way.
>
> —JAMES WATSON, *Genes, Girls, and Gamow*

> I don't like the word "big bang"; I never call it "big bang," because it is kind of cliché. This was invented, I think, by steady-state cosmologists.
>
> —GEORGE GAMOW, interview with Charles Weiner, 1968

RARELY DOES AN UNDERGRADUATE STUDENT END UP BECOMING A lifelong close friend and collaborator with his professor. But that's exactly what happened with Ralph Alpher, George Gamow's unlikely right-hand man. Alpher's association with his longtime mentor began in his undergraduate years at George Washington University (GWU) in the late 1930s and lasted until Gamow's death in 1968. Whereas Friedmann and Lemaître had derived the mathematical basis for the

Big Bang (as Hoyle had dubbed it), Alpher worked with Gamow and researcher Robert Herman to develop it into a physically realistic model that offered important insights into how matter and energy developed in the early universe.

Alpher's and Gamow's life adventures couldn't have contrasted more. Alpher was born and raised in Washington, DC, unlike his mentor, a Soviet émigré who had resided in many parts of the world. Whereas Gamow's physics education would have been on the fast track if it weren't for political developments, Alpher was forced to proceed cautiously for financial reasons, working at an assortment of jobs to bolster his income. He had the patience for detailed calculations that Gamow lacked. Though he greatly valued Gamow's intellect and generally enjoyed his mentor's jokes, he sometimes worried that zany "Gamovian" antics would overshadow their message. For example, Gamow often pointed out the similarity of their names to the Greek letters alpha and gamma, which, though amusing, made Alpher's contributions seem a bit like a punchline. The fact that one of their major joint papers was published (through sheer coincidence) on April Fool's Day didn't help matters.

Undoubtedly, in their collaborative efforts, Gamow was always the center of attention. In his early career, Alpher seemed to accept that. He found it to be a problem, however, later in life (after Gamow died) when he realized that he and Herman had not been allotted proper credit for their independent achievements, including their uncannily close prediction of the temperature of the relic radiation in the universe, confirmed in the mid-1960s. Once a loyal apprentice kept out of the spotlight, Alpher rightly argued that he deserved proper recognition as an accomplished scientist in his own right.

THE SORCERER'S APPRENTICE

If a group of tourists were to visit the Washington Monument in the nation's capital sometime in the summer of 1935 or 1936, they might have stopped to visit an open-air theater set up on its grounds. There,

American physicist Ralph Alpher, who studied under and collaborated with Gamow. He and Robert Herman correctly predicted the existence of cosmic microwave background radiation. CREDIT: AIP Emilio Segrè Visual Archives, Physics Today Collection.

if they happened to look behind the scenes of the performance, they might have noticed a rather ambitious teenager helping out with the props and staging. For young Ralph Alpher, then a student at a business high school in DC, the stage crew job offered him a steady wage and the chance to see fun plays and musicals. He knew that his backstage efforts were essential to putting on a fantastic show.

Upon high school graduation in 1937, when he was only sixteen, Alpher wasn't sure what to do. Adept at shorthand and typing, he found a position as secretary for the National Cash Register Company, and worked there for a brief time before enrolling at Woodrow Wilson, a free teacher's college. Discovering that the coursework there wasn't challenging, he changed his mind again and left after only a couple of months.

As luck would have it, in February 1938 a local scientific establishment was looking for a secretary, so Alpher applied. This place, the Department of Terrestrial Magnetism (DTM) of the Carnegie Institution in Washington, was the one that had recruited Gamow. Alpher's job offered him the splendid opportunity to become immersed in the groundbreaking nuclear research Tuve and Hafstad conducted there using the lab's six-hundred-thousand-volt Van de Graaff devices.

At the same time, Alpher began studying part-time at GWU. There, he was thrilled to take a course in relativity with Gamow, whom he admired because of his popular works. Gamow proved an extraordinary professor—and a bit of a showman—as Alpher recalled: "I found him a tremendously stimulating person. And he obviously loved physics, and enjoyed it, and he conveyed to me as a student a sense of enthusiasm which was hard to ignore."[1]

A memorable instructor, somewhat of a ham, Gamow loved to wow his students with baffling demonstrations, even if they didn't always go as planned, to the class's amusement. Once (later in his teaching career) he tried to reproduce Archimedes's famous experiment to determine whether a crown was truly made of gold. Archimedes, an ancient Greek mathematician and physicist, during one of his baths suddenly realized that the solution to the problem was water immersion—which compelled him to jump out of the tub and run outside shouting "Eureka!"

In Gamow's rendition of demonstrating the buoyant force, he set up the experiment with a crown (made of bronze) suspended above a giant water-filled beaker by a string attached to a stand. He tried to lower the crown gently into the water, but it suddenly came loose, smashed into the beaker, cracked the glass, and soaked Gamow and the students in the front row. The whole class erupted in hysterics.

The drenched Gamow, undaunted, walked over to the classroom's sink and announced that he was going to fill the basin and retry the experiment. Haplessly, he turned the wrong faucet and steam jetted out instead of water, fogging up his glasses. He couldn't see to turn it off and spent some time blindly searching for the knob before he could stop the steam. The class laughed even harder. He tried a second faucet, which he didn't realize was connected to a long hose. Water came out, but instead of flowing into the sink, it poured relentlessly all over the classroom, making the mess even worse. Finally, when his lab assistant rushed to clean up the disaster, Gamow was forced to concede that the equipment had defeated him.[2]

Alpher loved Gamow's zeal for physics and latched onto him like an eager rookie admiring a seasoned athlete. He attended some of the Washington Conferences that Gamow and Teller set up, including the famous 1939 meeting at which Bohr announced the German nuclear fission breakthrough. Realizing the importance of that announcement, Alpher was amazed by how quickly Tuve and Hafstad set up their own successful uranium fission experiment at the DTM.[3]

When the United States entered World War II at the end of 1941, Alpher decided to remain in or near Washington rather than seeking a position at one of the Manhattan Project's major research centers, such as Los Alamos. He didn't want to abandon his studies at GWU. Soon, an opportunity arose for him to work at the Applied Physics Lab (APL), a new center for military research associated with Johns Hopkins University in Baltimore. Because of its proximity, he was able to serve as an engineer there in charge of producing fuses for torpedo exploders, while conducting a master's thesis research project under Gamow's supervision on the topic of how stars produce their energy through nuclear fusion. In 1945, the year the war ended, Alpher completed his master's degree.

THE NEUTRON DANCE

After the war, Alpher continued to work full-time at APL. Conveniently, around that time, Gamow began serving at a consultant for that lab, which enabled the two men to meet, sometimes in Alpher's office, to chat about physics. Alpher's burgeoning fascination with astrophysics drove him to do the unfathomable: maintain his strenuous daytime position while enrolling in GWU's challenging PhD program, once again under Gamow's supervision. He managed to squeeze in the necessary coursework, completing his homework and research projects in the evening.[4]

Alpher's focus shifted to an intriguing research question concerning whether or not, according to the general theory of relativity, perturbations (stretches or dents) in an expanding substance would grow or shrink over time. In other words, if you were blowing up a balloon and stretched a section of it into an extra bump while it continued growing, would the protuberance grow even bigger or smooth out? One possible application was to see whether small globs of mass might develop into the seeds of stars and other astral bodies as the universe expands.

Alpher spent about a year working on that topic and was about to write up his results when one day Gamow stopped by his office waving a Soviet journal. Gamow showed him an article by the Russian physicist Evgeny M. Lifshitz (an associate of Gamow's good friend Landau) that examined the very question Alpher was working on. Lifshitz had published similar conclusions about the behavior of perturbations. Alpher realized, to his horror, that his research topic had been milked dry, and there was nothing more to say. Despondent that he had wasted all that time, he grabbed all his notes on the obsolete thesis project, ripped them to shreds, and flushed them down the toilet.[5]

A few days later, once he had calmed down, Alpher asked Gamow to help him find a new project. Gamow pointed to preliminary research he had completed about the origin of all of the chemical elements in the Big Bang. His hunch was that only the fiery, ultradense primordial cosmos, with temperatures in the billions of degrees, would offer the perfect furnace for all the types of atomic nuclei, from helium to uranium, to be forged. The process would need to be "nonequilibrium," meaning that, in contrast to the equilibrium found in the cores of vibrant stars, conditions such as temperature and density would have been rapidly changing. He had given a talk on the subject in 1942 to the Washington Academy of Sciences and had published a letter sketching his ideas in a 1946 issue of *Physical Review*. Perhaps Alpher, he suggested, could flesh out the concept by developing a theoretical model and use it to calculate the quantities of each element that could have been produced in the Big Bang. Those theoretical estimates could then be compared to the actual known amounts of each element.

Agreeing to the project, Alpher set to work under Gamow's direction on a simple model of element building called neutron capture. The idea was that the universe began as a thick, hot broth of radiation, sprinkled through with neutrons, which Alpher dubbed "ylem." During the first few minutes of creation, as the cosmos grew, some of the neutrons within the ylem underwent beta decay into protons and electrons. Some of the protons then captured neutrons in their vicinity (by means of nuclear forces) and became deuterons (proton–neutron combinations). Neutrons combined with deuterons to form helium-3 nuclei. Those captured more neutrons, which decayed into protons and electrons, resulting in helium-4 (plus electrons). That process, according to Alpher's thesis, kept going until all the chemical elements formed. It needed to happen very quickly because within minutes the universe would have expanded and cooled down so much that further element creation would be impossible.

Alpher was lucky to have the perfect sounding board for his ideas: Robert "Bob" Herman, a colleague at APL. They had met in the final year of the war and had become good friends. Herman, who had earned a PhD at Princeton in 1940 (the topic was molecular spectroscopy), was knowledgeable about relativity and cosmology. During his course of study, he had taken a class with Robertson, who had brilliantly cataloged the types of homogeneous, isotropic cosmologies that Einstein's equations allowed for. Robertson also chaired Herman's dissertation defense committee.

At APL, Herman had been engaged in the top-secret task of developing a new system for triggering antiaircraft rounds and other propelled explosives to detonate when they neared their targets, as gauged by signals sent and received from miniature radios embedded within the shells. That technology, called the variable time (VT) proximity fuse, had been designed in part by space scientist James Van Allen, who also worked at APL. After the war, Van Allen switched to the study of cosmic rays in the atmosphere, a project to which Alpher and Herman contributed. (In 1958, Van Allen would make his most famous discovery: the zones of charged particles in space captured by Earth's magnetic field, dubbed the "Van Allen belts.")

By 1948, Herman would join Alpher, and sometimes Alpher and Gamow, in numerous research collaborations related to Big Bang nucleosynthesis. Gamow greatly enjoying joking around with both of them—often acting like a teasing older brother. When Alpher and Gamow completed their first joint paper on the topic (assisted by Herman, who would be listed as an author on later papers), they celebrated by altering the label of a bottle of Cointreau liqueur to read YLEM. (Alpher would eventually donate the bottle to the Smithsonian Institution, where it has been displayed in the Explore the Universe gallery.[6])

ALPHA, BETA, GAMMA, AND SOMETIMES DELTA

While Alpher was excited when Gamow submitted their joint paper on Big Bang nucleosynthesis, entitled "The Origin of Chemical Elements," for publication, he was taken aback by the inclusion of a third author, Hans Bethe, who had not been involved at all. The sole purpose of listing Bethe's name, noted as in absentia, was to render the authorship list, arranged as Alpher, Bethe, and Gamow, a pun on the first three Greek letters. The article would come to be known as the "alpha, beta, gamma" paper.

Bethe, who was grateful to Gamow for inviting him to the Washington Conferences that helped spark his important research in stellar nuclear physics, and who was amused by Gamow's pun on their names, didn't complain about the inclusion—even if it was strange. Many physicists with bigger egos would have balked at the idea of someone putting their name on a speculative article they had nothing to do with without their permission. But Bethe, who was immersed in his own attribution dispute (Victor "Viki" Weisskopf believed that Bethe should have given him more credit for certain ideas about the quantum properties of electrons), remained cordial. Invited by Gamow, he agreed to serve on Alpher's dissertation defense committee.

However, Alpher took the paper—mainly some of the results of his thesis project—very seriously and was miffed to have to share credit with a physicist who hadn't contributed to the project. He had worked long and hard on the calculations, staying up late at night to do so while maintaining high standards in his full-time job during the day, whereas Bethe hadn't lifted a finger. But Gamow was his supervisor, so what could he do?

Furthermore, as Alpher perceptively noted, Gamow's jovial approach had the potential of alienating scientists who took their work more seriously. Indeed, as Gamow's career progressed, his reputation as a jokester and humorous teller of tales began to overshadow his eminence as a brilliant theorist. His drinking habit certainly didn't help matters. Alpher would work hard to get the word out about their research in a more solemn way, pointing out its prescience and importance.

The key triumph of the "alpha, beta, gamma" paper is that it was the first to deduce—with reasonable accuracy—why the universe contains so much helium. Stellar models couldn't account for the abundance of that element. Only the supposition that it was forged in a hot primordial fireball produced a solid picture of how it emerged.

Pleased by the results, Gamow sent off copies to several physicists, including Einstein and Swedish physicist Oskar Klein, whom he had come to know during his time at Bohr's institute. In his response to Gamow, Einstein was very positive. He commended the mechanism for element creation described in the paper.[7]

In the copy for Klein, Gamow wrote in the cover letter: "It seems that this 'alphabetical' article may represent alpha to omega of the element production. How do you like it?"[8]

Klein replied: "Thank you very much for sending me your charming alphabetical paper. Will you allow me, however, to have some doubt as to its representing 'the alpha to omega of the element production.' As far as gamma goes, I agree of course completely with you and that this bright beginning looks most promising indeed, but as to the further development I see difficulties."[9]

It was apt of Klein to point out difficulties with the paper's model of how the higher elements (beyond helium) were produced. As Princeton physicist and Nobel laureate James Peebles would note: "Gamow was a genius at physical intuition. He was never interested in details. In those 1948 papers were startlingly inventive and beautiful physics. He thought that possibly all the elements could be created in the hot big bang. They showed very clearly that you'll get a lot of helium. They got a lot of helium and not much beyond that."[10]

Stanford physicist Robert Wagoner, who in the 1960s would join with Hoyle and Fowler in calculating the helium abundance more accurately, pointed out a major problem with Gamow's (and his group's) methodology: "He motivated physicists and astronomers to investigate the early universe. However, he made a serious error by including only the decay of the neutron, not the more important reactions between neutrons, protons, neutrinos, and electrons (and positrons) which determine the subsequent formation of the lightest elements."[11]

By the time of the "alpha, beta, gamma" paper's release—on April 1, 1948—Herman had fully joined the project as a collaborator. For a follow-up paper in which he was included, Gamow teased him that he should change his name to "Delter" to continue the pun on Greek letters. Gamow would later rib Herman that the first syllable of his surname, "Her," sounded similar to "kher" (a vulgar expression in Russian for "penis"). Therefore, he joked, Herman should at least go by "Deltman."[12] Herman took such comments in stride, knowing their unrestrained source.

The grand scope of the researchers' work attracted press coverage. On April 14, the *Washington Post* published an article about the rapid pace of Big Bang nucleosynthesis, under the startling headline "World Began in 5 Minutes, New Theory." Two days later, the celebrated cartoonist Herbert

Block, professionally known as "Herblock," released in the same paper a cartoon depicting an atomic bomb in a pensive pose reading that news item and muttering to itself, "Five minutes, eh?"

GLIMMERS OF A COSMIC GLOW

Throughout 1948, in preparation for his thesis defense, Alpher worked with Herman to refine the Big Bang nucleosynthesis model. They were startled by how quickly Gamow was releasing papers before they had time to double-check the results. His impulsivity was getting the better of him. Ambitiously, he had completed some rough calculations on his own about early conditions in the universe that might lead to galaxy formation and then hastily sent off his results to the leading science journal *Nature* as a single-author work. When Alpher and Herman looked over Gamow's work, they found major computational errors, but it was too late for Gamow to make corrections. Therefore, with his permission, they submitted their own article with their own calculations to *Nature*, indicating the points in Gamow's paper that it rectified.

In that work and in a follow-up paper, Alpher and Herman made an astonishingly prescient prediction about the lingering afterglow of the Big Bang. Tracing the hot early radiation as it cooled down over time, they predicted the temperature of space in the current era would be 5 K (Kelvin, or degrees above absolute zero). Penzias and Wilson, in their 1965 paper, find the value to be around 3 K, so Alpher and Herman weren't far off, as it turned out.

Gamow, who would later make his own (higher-temperature) predictions about cosmic background radiation, advised Alpher at the time that the 5 K prediction was essentially meaningless. For some reason Gamow wrongly surmised that the combined effect of shining stars would bathe space in a 5 K glow that would drown out any signs of leftover radiation from the Big Bang any cooler than that. During the summer of 1948 Gamow was in Los Alamos at the invitation of Teller to complete some military research. He wrote to Alpher: "The only thing we can tell is that the 'residual temperature' of the original heat of the universe *is not higher* than 5 K, but it could be as close to zero as one likes."[13]

Arguably, Gamow should have been more patient in reviewing Alpher and Herman's results before dismissing their importance. His penchant for rendering hasty judgments and then moving on was the downside of his seat-of-the-pants approach to physics. If the three of them had presented a unified message in the late 1940s about a cooled-down cosmic radiation background, perhaps experimental researchers would have tested that hypothesis sooner. Rather, a serendipitous discovery was matched with the prediction only in hindsight.

On March 18, 1949, Alpher was invited to explain the idea of element production in the hot early universe on the *Johns Hopkins Science Review* weekly television show. Broadcast to hundreds of thousands of viewers, many of whom had just bought their first television set, the show helped him get the word out about Big Bang nucleosynthesis. As he described on the show (referring to a cosmic age estimate that we now know to be far too low):

> About 3 billion years ago, after the universe began to expand, conditions were so hot that atoms boiled down to their elementary particles. After about four minutes of expansion, the universe cooled down to 1,000 million degrees, and protons formed from the radioactivity of neutrons, then stuck with the neutrons to form heavy hydrogen. This started the process of element building. The heavy hydrogen atoms in turn captured neutrons to form still heavier hydrogen; the next successive capture led to helium, and so on up through the heaviest elements such as uranium. This theory fits the observed relative abundance data as we see here. The variation in relative abundance from the light to the heavy elements is due to the fact that the ability of atoms to capture neutrons varies from light to heavy elements in a systematic way.[14]

Unfortunately, Alpher's vision was soon proven misguided. Around late October 1949, he and Herman received a manuscript and cover letter from Anthony "Tony" Turkevich of the University of Chicago with troubling news about the prospects of Big Bang nucleosynthesis of the higher elements beyond lithium. Along with eminent Italian émigré physicist Enrico Fermi, Turkevich had systematically investigated the nuclear processes proposed in the papers of Alpher, Herman, and Gamow and found a major stumbling block.

According to Fermi and Turkevich, in their unpublished manuscript "Calculations on the Formation of Light Nuclei on the Gamov [*sic*] Scheme," the fact that no stable isotopes exist with the atomic mass number (an element's number of neutrons plus protons) of 5 or 8 meant that it would have been impossible for almost all of the heavy elements to build up during the Big Bang using the neutron capture method.[15] Lithium-5 (with three protons and two neutrons) decays in about 10^{-24} seconds and beryllium-8 (with four protons and four neutrons) in about 10^{-17} seconds, making each so unstable that it could not possibly capture single neutrons or protons before disintegrating. Therefore, anything more complex could not be fused together in the fiery minutes of the nascent expanding universe.

An analogy can help illustrate how the neutron capture method fails to produce higher elements. Imagine climbing a wooden ladder from isotope to isotope, with each ladder rung representing the addition of a neutron or a proton. Suppose, on such a ladder, rungs 5 and 8 were pocked with termite damage such that the merest weight on them would cause them to collapse. In that case, no one would be able to ascend the ladder step-by-step. Similarly, simple nuclear processes could not lead to higher elements such as carbon, oxygen, and others beyond the rotten rungs on the ladder.

The good news conveyed in the Fermi-Turkevich result, however, was that Big Bang nucleosynthesis seemed to successfully explain the large fraction of helium in the universe (around 25 percent of its total elemental content) in a way that stellar models could not. Therefore,

Gamow, Alpher, and Herman could revel in that at least one important prediction of their model would stand.

In 1950, Alpher and Herman published a joint paper, "Theory of the Origin and Relative Abundance Distribution of the Elements," referencing, explaining, and responding to the Fermi-Turkevich result. They would continue to look for a way around the missing rungs of the ladder. Meanwhile, a running joke in the astrophysics community, which ironically turned out to actually be true, was "Gamow's theory is a wonderful way to build up the elements all the way to helium."[16]

THE POPE'S BIG BANG BLESSING

At that point, Gamow distanced himself, in his workload, from Big Bang cosmology. He left the tricky calculations to Alpher and Herman and turned his thinking to other projects. Aside from releasing his popular book, *Creation of the Universe*, he harbored the frontiersman's itch to gather the wagons and head off to greener territory.

But his Big Bang theory received an unexpected boost from an unlikely locale: the Vatican. Pope Pius XII, who had a passionate interest in astronomy, had learned much about the theory of a finite universe and was convinced that it had a deep connection to the biblical Genesis. Catholic faith allows the possibility of novel interpretations of divine truth as humanity develops new ways of understanding the world. Astronomy's findings are particularly welcome if they support the wonder and eminence of God. Accordingly, the pope began to argue that divine creation was borne out by modern theories, albeit on a far greater scale than the ancients believed.

Lemaître had been a member of the Pontifical Academy. Though a devout Catholic, he took issue with the pope's interpretation. Behind the scenes, he argued that astrophysical discussions should remain separate from matters of faith. The pope listened intently but nevertheless remained steady in his views on cosmology.

The pope first announced his opinion of the reconciliation of Big Bang cosmology with biblical teachings at a meeting of the Pontifical Academy on November 22, 1951. (Traveling at the time, Lemaître wasn't present.) He remarked that "everything seems to indicate that the material universe has had, in finite time, a powerful start, provided as it was with an unimaginable abundance of reserves in energy; then, with increasing slowness, it has evolved to its present state."[17]

Gamow, though not a Catholic or even religious, was delighted to learn about the pope's endorsement. He soon mailed the pope a copy of a popular article he had written about Big Bang cosmology. When *The Creation of the Universe* was released, Gamow sent the pope a copy of that, too.

At a 1952 meeting of the International Astronomical Union, hosted at his summer residence, Castel Gandolfo, the pope told attendees: "Such mysteries [of cosmology] postulate and point to the existence of one spirit that is infinitely superior, the divine creative spirit Who creates everything that exists, conserves it in being and governs it, and meanwhile, with supreme insight, knows and scrutinizes His handiwork, just as he did on the first day of creation."

The *New York Times* reported that speech under the eyebrow-raising banner "Pope Says Science Proves God Exists."[18]

Despite the papal blessing, supporters of the Big Bang continued to grapple with one major issue. Data gathered with various telescopes—including the Hale Telescope at Palomar Observatory in San Diego, which at the time had the largest mirror in the world—supported an age of the universe that was too brief. Until the age of the cosmos could be shown to be greater than that of the stars and galaxies, the steady-state theory remained a reasonable alternative.

In 1953, Gamow found an exciting new pursuit: genetics. Stimulated by the famous paper by James Watson and Francis Crick describing the double helix structure of DNA on which patterns of four bases (chemical ingredients) were arranged along parallel strands, he began to wonder how the specific order of those bases was used to form the proteins needed for life. By that time, biologists had identified twenty distinct amino acids needed in various combinations to create known proteins. Familiar with probability theory, Gamow thought about how four different "digits" (associating the bases with numbers) could be arranged to create the amino acids in the proper sequence to generate the proteins. He wrote to Watson and Crick in July of that year with his ideas.

Watson was impressed by Gamow's suggestion to apply combinatorics to genetics. The next time he was in the DC area, he arranged a meeting with Gamow at his house in Bethesda. As soon as Watson arrived, Gamow handed him a copy of the Japanese translation of his most recent book, *Mr. Tompkins Learns the Facts of Life*, which happened to be on the subject of biology. When Watson opened the front cover to look for an inscription, he was pranked by the message "I've fooled you—open the other side." Gamow chuckled that his practical joke worked; Watson had not known that Japanese books opened in the opposite fashion to those in English.

Gamow was very proud that his books were translated into so many languages and available internationally. According to Princeton astrophysicist J. Richard Gott III, "In his basement, Gamow had a whole wall of books, all editions of his books in different languages."[19]

After playing his prank, Gamow handed Watson a scotch and soda, and a new friendship was born. They'd remain in close touch for years, as Gamow started to publish in biology, proposing a "triplet code" to explain how amino acids emerged from the pairs of bases. His mathematical concept helped lead to a full understanding of the genetic code: linking patterns of the nucleotides in DNA and RNA with the formation of particular kinds of proteins requisite for life.

As Gamow's son, Igor, recalled: "He and Father had an interesting relationship. Watson was fascinated with physics. He was pumping Father with the question, 'What is life?'"[20]

Along with Watson, Gamow assembled a group called the "RNA Tie Club," consisting of various scientists solving the riddles of genetics.

He limited the club to precisely twenty members, each nicknamed after one of the amino acids. Gamow's was "Alanine." Each member received a special wool tie, embroidered with an RNA helix. Although ultimately Gamow didn't solve the puzzle of genetics, he made important suggestions—such as the role of mathematical combinations of bases in encoding the various types of amino acids—consistent with his penchant for sparking ideas, but not necessarily following them to completion.

FAMILY MATTERS

Igor recalled that amid his great bursts of creativity, Geo Gamow remained an attentive father. When Igor was a child, upon hearing his dad's stories about motorcycling through Europe, he decided that he wanted a motorcycle for himself. His father was happy to comply.

"Father researched and found a German motorcycle," Igor recalled. "I was the first kid in junior high school to go to school on a motorcycle, and I became that hated person in the neighborhood. All my life I have been an easy rider."[21]

Geo and Rho soon were forced to deal with troubling issues pertaining to Igor's education. They all spoke Russian at home, resulting in a delay in Igor's learning English. Staff at the public school he was attending in Bethesda recommended that he be sent to a boarding school, separate from the family, to improve his English-language skills. Consequently, from age twelve to sixteen, Igor attended a military academy in Staunton, Virginia. Finally, he decided to leave high school, with thoughts of becoming a cowboy.

After some time, Igor postponed that idea and instead decided to join the ballet. Triumphantly, he joined a company and was soloist in a production of The Nutcracker. Unfortunately, he found himself short of breath and coughing a lot during performances—which he attributed to his parents' excessive smoking—and had to give up dancing. As he recalled: "Father went to a performance and said, 'You know, Igor, maybe I should bring you an oxygen tank.'"[22]

Igor would later embark on a successful career in microbiology and inventing. To finally fulfill his cowboy dream, he also bought a ranch in Colorado and raised horses as a sideline. But sadly, well before those successes, he would witness the dissolution of his parents' marriage.

Although they continued to share intellectual interests, Geo and Rho were an emotional powder keg. Their once-happy marriage had turned into a succession of bitter arguments. Finally, in 1956, they filed for divorce. Around the same time, Geo decided to leave GWU. Social life there was very close-knit, which brought him often into contact with President Marvin, who was reportedly rather religious and disapproving of divorce. (There are unconfirmed speculations—for example, by Alpher—that Marvin asked Gamow to leave.)[23]

After GWU, Gamow spent some time at the Ohio State University before finding a position in the physics department at the University of Colorado, Boulder. In 1958, after falling in love with Barbara "Perky" Perkins—an editor and translator at Cambridge University Press—he'd get remarried. They then lived in a house he called the "Gamow Dacha." (*Dacha* is the Russian word for country cottage.) Together with Barbara, he'd remain in Boulder the rest of his life, enjoying Colorado's spectacular rocky scenery through numerous car trips to delightful places such as the Garden of the Gods.[24]

"I like the pioneering thing," Gamow explained. "I like [the Colorado] mountains much better than [those in] California, where they have a hotdog stand on the top of each mountain."[25]

Coincidentally, Gamow moved out west at a time when Hoyle was spending much time in the western part of the United States as well. On his 1953 research visit to Caltech, Hoyle had forged important connections. Through Fowler, he formed a close association with the Kellogg Lab, which would prove an ideal place to test his theories of the creation of the heavy chemical elements in the collapsed cores of dying stars. Through that work, he'd complete Gamow's mission of solving the problem of element building beyond helium, albeit in a manner far different from Big Bang nucleosynthesis.

The two frontier thinkers would finally meet in person during the summer of 1956, when both happened to be in California. Gamow was serving for two months as a consultant to General Dynamics, an aeronautics and defense company in La Jolla, a seaside neighborhood of San Diego. At the same time, Hoyle was in Pasadena—about 120 miles to the northwest—conducting research at Caltech while on leave from Cambridge. Upon learning of their relative proximity, Gamow contacted Hoyle and invited him down for a visit.

Thanks to his lucrative summer job, Gamow was able to afford a cushy white Cadillac, with which he proudly drove Hoyle around La Jolla. They could finally debate—in a friendly way—their opposing views of the universe. Among their topics of discussion during the car ride was the temperature of space. By that point, Gamow had settled on a value of about 7 K, due to cooled-down relic radiation from the Big Bang—somewhat higher than Alpher and Herman's earlier estimate. Hoyle insisted it was about 0 K (close to absolute zero), based on a steady-state scenario involving the continuous creation of small quantities of matter in space.

Hoyle brought up the research of Canadian astronomer Andrew McKellar, who in 1941 had estimated the temperature of space to be about 2 to 3 K. The basis of McKellar's argument was not cosmological, but, rather, having to do with the spectral lines of simple molecules such as cyanogen (single atoms of carbon and nitrogen bonded together) and methyne (single atoms of carbon and hydrogen bonded together). Such spectra indicated that the molecules were bathed in incredibly frigid radiation of only a few degrees above absolute zero. Hoyle pointed to that result to rule out Gamow's higher estimate of the background radiation temperature. Not surprisingly, given their cosmological debate, the two thinkers did not come to an agreement.

Years later, Hoyle regretted that the two had not settled on a value of around 3 K, matching the result of Penzias and Wilson published in 1965. Perhaps, he thought, they would have been hailed as predictors of the cosmic microwave background radiation discovery. "Whether it was the too-great comfort of the Cadillac," Hoyle wrote, "or because George wanted a temperature higher than 3 K, whereas I wanted a temperature of zero K, we missed the chance of spotting the discovery made nine years later by Arno Penzias and Bob Wilson. For my sins, I missed it again in exactly the same way in a discussion with Bob Dicke at the 20th Varenna summer school on relativity in 1961."[26]

CHAPTER SIX

Building the Elements

> I believe my father is the only scientist ever to have explained why
> he believe something to be true by using a philosophy argument,
> namely when asked by Caltech why it should bother to look
> for Carbon-12 when it didn't exist, he replied: "I exist, therefore,
> Carbon-12 exists."
>
> —Elizabeth Jeanne Hoyle Butler

> In the beginning God created radiation and ylem.... And God said:
> "Let there be mass two." And there was mass two. And God saw
> deuterium and it was good....
>
> And God continued to call number after number.... In the
> excitement of counting, He missed calling for mass five and so,
> naturally, no heavier elements could have been formed....
>
> And God said "Let there be Hoyle." And there was Hoyle. And God
> looked at Hoyle... and told him to make heavy elements in any way
> he pleased.
>
> And Hoyle decided to make heavy elements in stars, and to spread
> them around by supernovae explosions....
>
> And so, with the help of God, Hoyle made heavy elements in this
> way, but it was so complicated that nowadays neither Hoyle, nor
> God, nor anybody else can figure out exactly how it was done.
>
> —George Gamow, "New Genesis," in *My World Line*

Sunny Pasadena, home of Caltech, has a reputation for inspiring genius. Einstein's visits to that Southern California town in the early 1930s as a Caltech visiting researcher led to his fruitful discourse with Hubble, Lemaître, and other notable scientists about the nature of the universe. Max Delbrück, a trained physicist (and friend of Gamow) who became a molecular biologist, conducted much of his Nobel Prize–winning studies of bacteriophages (viruses that infect and replicate within bacteria) there.[1] Linus Pauling, Richard Feynman, and so many other eminent researchers thrived there. Robert W. Wilson would earn his PhD there before codiscovering the telltale hiss of the cosmic microwave background radiation. The list goes on and on.

For Hoyle, Pasadena represented a haven of academic freedom away from stuffy Cambridge life—a place where he could freely engage in his research and, as it turned out, make some of his most significant breakthroughs. He loved Great Britain for its unassuming people in country towns and rugged walking paths that wound past lakes and over moors, hills, and dales, but he hated British bureaucracy and hierarchy. Decisions such as funding, resource sharing, and appointments seemed to be based on status and pedigree rather than actual merit. As his daughter, Elizabeth, recalled: "He was very un-self-important, and the only people who tended to rub him up the wrong way were the self-important people who were both what we here in the UK call upper middle class, and believed they were entitled to respect they didn't actually earn."[2]

Hoyle agreed with his friend and mentor Ray Lyttleton when, in a famous squabble at the Royal Astronomical Society, Lyttleton suggested to its president that its journal needed competent referees. After the president assured him vociferously that eminent astronomers review all the papers, Lyttleton had snapped back, "Sir, you must have misheard me. I did not say eminent, I said competent."[3]

In Hoyle's mind, the United States seemed relatively free of such pretensions. The idea, not the social status of its proposer, seemed to be what was important. The American university scene, particularly in

Southern California, seemed a kind of nirvana for gathering informally, rolling up one's shirtsleeves, and discussing revolutionary new concepts.

Meeting Willy Fowler in 1953 at his house party after Hoyle's talk at Caltech on the expanding universe offered such an opportunity. Fowler was the antithesis of a snobbish RAS official; rather, he loved to chat about sports and hobbies—from baseball and blackjack to steam trains—almost as much as talking about science. It was in Fowler's office in February that Hoyle raised vital questions about an energy state of carbon-12 that would change the course of astrophysics.

THE SWEET SPOT OF CARBON PRODUCTION

Although Hoyle hadn't written much about the subject since his 1946 paper, "The Synthesis of the Elements from Hydrogen," he had never given up hope of explaining how all the chemical elements emerged from stellar processes. The notion of Big Bang nucleosynthesis was far-fetched and antithetical to common sense, he believed. Why reach an extreme conclusion about something untestable that supposedly happened billions of years ago while the stars all around us were ripe for exploration? Not that he used this analogy, but he would have enjoyed it: It is like an accident-prone driver who, instead of learning how to read the traffic signals, blames an unknown distant ancestor for the current bad luck behind the wheel. Don't push something into the far past, Hoyle repeatedly argued, unless every single present-day option has been exhausted. Of course, if steady-state was right, there wasn't a Big Bang anyway, but that was a different argument.

Aside from a focus on the observable, another key argument for stellar nucleosynthesis had to do with the abundance of metals (elements beyond helium). Whereas the fractions of hydrogen and helium appeared to remain relatively steady over time, the quantities of all of the other elements seemed to have increased over the ages, timed with the development of massive stars in galaxies. Those facts suggested that the metals emerged via an ongoing process.[4]

Transforming hydrogen into helium via two different kinds of cycles (as described by Bethe) was by then well understood. The big hurdle for any form of nucleosynthesis of metals was "helium burning": the process of transforming helium into carbon to supply the fuel for even higher processes. The missing ladder rungs of isotopes with atomic mass numbers 5 and 8, pointed out by Fermi and Turkevich and others, greatly complicated the arithmetic.

Around 1952, Hoyle had begun to consider a "triple alpha process" in which three alpha particles, or helium nuclei, combined to form carbon-12. The idea was that two would come together, form the highly unstable isotope beryllium-8, and then fuse with a third alpha particle in the fleeting interval before the beryllium decayed, thereby forming carbon-12. Two challenges had to be met in solving the riddle. The first was explaining how all this could happen in such a brief instant. The second was justifying how the carbon that formed remained stable.

To Hoyle's chagrin, he found out that astrophysicist Edwin Salpeter had already solved the first problem. Salpeter proved that at temperatures above 100 million degrees Kelvin, the frequency of chance collisions with alpha particles would allow the possibility of a beryllium nucleus fusing with a particle before the beryllium disintegrated. Such high temperatures would be present, Hoyle realized, in the core of a red giant star, the blown-up state certain stars assume once they've exhausted their primary hydrogen fuel.

Fortunately for Hoyle, Salpeter hadn't addressed the second question, and Hoyle had a solid hunch about the solution. To ensure the likelihood and stability of the conversion of helium and beryllium to carbon-12, Hoyle postulated that carbon-12 had a special energy level of 7.65 MeV (million electron volts, a unit of energy) that had yet to be observed in the lab. That energy level would match the conditions of three alpha particles merging inside the core of a red giant. Quantum rules, in that case, would make such a transition go smoothly, increasing the chances of success by a factor of millions. Hence, this would explain how red giant cores could serve as "factories" for converting helium into carbon. Such carbon, once released into space with the demise of the star, would form an essential ingredient for life on Earth and potentially elsewhere. Thus, without such a process, we wouldn't be here. Later in life, Hoyle would tie this insight to the anthropic principle that the universe is a certain way because, if it were very different, intelligent life wouldn't have evolved. Therefore, the 7.65 MeV carbon resonance exists because we exist.

Hoyle consulted Fowler to see whether the idea might be testable. Fowler agreed to try to identify a 7.65 MeV carbon-12 resonance level by using his lab's powerful accelerator. Resonance is a surge at a particular energy level indicative of a transformation. A more familiar example is that when a skilled opera singer hits the frequency matching the resonance level of a glass, it vibrates wildly and perhaps even shatters.

Hoyle couldn't have asked for a better place to test his hypothesis. Founded by Danish nuclear physicist C. C. (Charles Christian) Lauritsen, the W. K. Kellogg Radiation Laboratory where Fowler worked had an outstanding record for exploring nuclear structure. Charles Christian's son Thomas Lauritsen, a noted physicist in his own right, also worked at the lab. Finally, along with Fowler and the two Lauritsens, another researcher, Charles W. Cook, worked at Kellogg in the mid-1950s. On Fowler's suggestion, and with the enthusiastic support of Ward Whaling, a Kellogg physicist who was an expert in nucleosynthesis, the four experimentalists sprang into action and soon, to their surprise and glee, discovered the carbon-12 resonance level Hoyle had predicted. Remarkably, it took them only around ten days to achieve that confirmation. Fowler simply couldn't believe that a brief conversation with Hoyle had led so quickly to such a stunning discovery.

When Cook, Fowler, and the Lauritsens published further results, they documented how Hoyle's prediction was almost miraculously on target. As they wrote: "It is concluded . . . that the second excited state of C12 as predicted by Hoyle is of a suitable character to act as a stellar thermal resonance in the Salpeter process . . . under conditions expected in red giant stars."[5]

Hoyle used the result to bolster a model of element production he had developed that explained how each element develops from simpler isotopes via a series of nuclear processes—starting with helium into carbon, and from there up to iron and nickel. He couldn't yet explain how even higher elements were formed but theorized that processes in even hotter cores or the intense energies released in rapid supernova bursts would do the trick. His seminal paper, "On Nuclear Reactions Occurring in Very Hot Stars: The Synthesis of Elements from Carbon to Nickel," was published in 1954.

Along with his intense research, Hoyle took time to explore the wild landscapes of the United States. During his visit, he had the pleasure of hiking the Grand Canyon and visiting other spectacular sights. He also became acquainted with notable scientists, particularly Hubble, who sadly died later that year on September 28, 1953. After his death, the Hoyle family would get to know his wife, Grace, during subsequent visits to Caltech. As Hoyle's son, Geoffrey, recalled: "I sometimes used to stay with Mrs. Hubble when in Pasadena and she said my father would listen patiently to her husband even if he disagreed with what was being said."[6]

Hoyle's first stay at Caltech, though only a few months long, proved extremely productive. He needed to return to Cambridge to teach his classes, but he had been bitten by the American bug and had a deep desire to return and complete his project. That quest would lead him to form an important team of astrophysical researchers that would come to be known as B^{2}FH.

A STELLAR QUARTET

Musical quartets harmonize in a manner that often transcends the contributions of the individual performers. Similarly, the scientific duo of Hoyle and Fowler benefited by inviting Margaret (née Eleanor Margaret Peachey) and Geoffrey "Geoff" Burbidge, a wife and husband team of astronomers, to join their ensemble. (Hence the "B squared" in B^{2}FH.)

The four meshed seamlessly, creating glorious insights about the chemical ladder.

Margaret and Geoff were a powerful couple, each strong-willed and persistent. In appearance, they offered a stark contrast. She was petite and soft-spoken; he was hulking and loud. Yet, in an age of rampant sexism, when women scientists were often slighted, they saw each other as equals and worked together to fight discrimination. Both from England, they had met at University College London, where she was studying astronomy and he was studying physics. They married in 1948.

A summer visit to Yerkes Observatory in 1952 brought the couple face-to-face with the question of how the chemical elements arose. Acting as a kind of reporter, Margaret attended a conference there on element formation, with the aim of taking careful notes and writing up the proceedings in *Observatory Magazine*. Geoff went along with her, and they both found the experience eye-opening. The slant of the meeting was Big Bang nucleosynthesis, focusing on the Gamow, Alpher, and Herman scheme, as well as an alternative idea put forth in 1949 by nuclear physicist Maria Goeppert-Mayer and Edward Teller that involved the fission (breaking up) of a cold, primordial nuclear fluid packed with neutrons into various heavy elements. The latter, known as the "polyneutron" theory, was shown that summer by Peierls and his collaborators to have a fatal flaw: it predicted, against all experience, that there would be the same (or higher) proportion of heavy elements in the universe as lighter elements. Gamow and Mayer each gave talks at the conference on their respective theories of Big Bang nucleosynthesis (hot versus cold). Hoyle's 1946 paper on stellar nucleosynthesis was mentioned only briefly. Still, the conference gave the Burbidges a taste of an exciting new enterprise. They attended a conference in Michigan the following summer, including talks by Baade and Gamow, which piqued their interest even further.[7]

At the time, many observatories in Britain did not offer women telescope time, a serious hindrance for Margaret. Though the situation was not much better in the United States, the relative flexibility of the

American social order offered greater hope that observatories would soon be more welcoming to women. Therefore, the Burbidges kept an eye out for opportunities in the USA. Soon their wish was fulfilled. In the meanwhile, they moved to Cambridge, with Geoff supported by a research fellowship from Imperial Chemical Industries.

In fall 1953, Fowler arrived for a yearlong sabbatical leave in Cambridge and immediately hit it off with the Burbidges. The three had much to discuss about spectral analyses of stars, the rainbow patterns of lines of various colors that indicated the presence of chemical elements in stellar atmospheres. They were looking for signals of unusual amounts of heavy elements that might point to signs of novel physics, such as processes involving neutron capture by various nuclei that transformed them into higher elements. Frustratingly, Hoyle was so busy teaching at that point—making up for his own leave—he barely had time to meet with Fowler, get to know the Burbidges, or listen to their ideas about stellar processes, let alone work with them. Fowler returned to Caltech at the end of spring 1954, leaving Hoyle stewing about a missed opportunity.

As a consolation for his hectic schedule, Hoyle was gratified, no doubt, by the growing support in Britain for the steady-state universe. On September 7, 1954, a meeting in Oxford of the British Association for the Advancement of Science essentially canonized the steady-state theory as the "new cosmology." Recall that the pope had already blessed the Big Bang. This was like the English religious schism of the sixteenth century all over again, only for cosmology rather than theology. If someone was the "archbishop" of the British doctrine, it was the Astronomer Royal, Sir Harold Spencer Jones, who was an ardent believer in continuous creation. Hoyle, Bondi, and Gold were the leading interpreters and scribes of the faith. Finally, there was Sciama, the independent-minded disciple of steady-state who was always questioning but who remained in the flock as long as his queries were satisfied. In covering the event for the *New York Times*, reporter John Hillaby contrasted British support of

Margaret Burbidge, Geoffrey Burbidge, William Fowler, and Fred Hoyle, co-authors of the key B²FH paper explaining the origin of the chemical elements through stellar nucleosynthesis. Fowler, who was a great fan of steam trains, is enjoying a model Donald Clayton gave him as a gift for his sixtieth birthday. CREDIT: AIP Emilio Segrè Visual Archives, Clayton Collection.

steady-state with what he called the "American School of Dr. Gamow," which advocated that the universe began in "one initial explosion."[8]

Finally, Hoyle was offered the opportunity to return to Caltech, starting in January 1956 and lasting until June, in a visiting position called the Addison Green Whiteway Professorship, which had been arranged in connection with an unpaid temporary position at Mount Wilson Observatory. He got permission to return again in September and stay until the end of the year. That hiatus from Cambridge would prove the most fruitful period of his career. Auspiciously, the Burbidges were also in Caltech, with Margaret a visiting researcher at Kellogg Lab and Geoff supported by a Carnegie Fellowship. Along with Fowler, the B²FH quartet could finally assemble.

AN EXPLOSION OF RICHES

With Fowler, the Burbidges, and Hoyle all focused on neutron capture as a means of creating higher elements, they put their minds together to see how they could make it work. A visual aid turned out to be particularly useful: a chart showing each heavy isotope's atomic number (number of protons) versus atomic mass number, or nucleon number (number of protons plus neutrons). Like Goldilocks, only one combination of protons and neutrons for each element is just right for greatest stability. Greater or fewer neutrons than that number generally leads to isotopes that are less durable (and in worst cases, isotopes that are absolutely unstable and immediately decay). Therefore, there is an optimal atomic mass number that corresponds to each atomic number, and a sketch of this "stability line" (actually, a curve) for the elements runs roughly diagonally from the lower left to the upper right of the chart. The farther from the stability line, generally speaking, the more fragile the isotope.

The B^2FH team speculated that, for elements higher than iron, separate types of neutron capture-interactions produced isotopes to the left and the right of the stability line. Isotopes to the left were generally formed by what they called the "s process," standing for "slow neutrons," and those to the right by the "r process," standing for "rapid neutrons." In other words, neutron-poor isotopes typically evolve by capturing slow neutrons—which undergo beta decay to transform into protons (while emitting electrons and antineutrinos)—while neutron-rich isotopes generally develop by being bombarded with rapid neutrons that are captured repeatedly before they have time to decay. S processes, they surmised, take place within the core of certain types of red giant stars, whereas r processes occur in the tumult of certain kinds of supernova explosions. The team also proposed a p process, involving protons, but it ended up not proving significant.

This was a brilliant proposal, crafted by four highly capable individuals, to explain the world around us. As Fowler once remarked: "All of us are truly and literally a little bit of stardust."[9]

One might use an analogy to envision the difference between the results of slow and fast neutrons. Imagine a lemonade stand that sells "lemonade ice" (lemonade topped with crushed ice and served in a paper cone). The lemonade symbolizes the protons in a nucleus, and the crushed ice, the neutrons. The operator, a girl named Stella, serves three varieties: regular, with a mixture of about one-third lemonade and two-thirds ice; "watery," with watered-down lemonade; and "icy," with extra ice. If a customer orders a watery, then Stella places lemonade ice in a cone and slowly scoops bits of ice onto the serving so that the ice melts (symbolizing beta decay) and waters down the lemonade (symbolizing the addition of more protons in the s process). If a customer orders an icy, she fills a cone with lemonade ice and then rapidly scoops chunks of ice onto the serving (signifying the addition of more neutrons in the r process) so that the extra ice does not have time to melt. Similar to the mounting ice, fast bombardment allows neutrons to stay neutrons.

By September 1956, Hoyle needed to return to Cambridge and resume teaching. However, he was confident that the team's results were very worthy of publication and that the Burbidges, he, and Fowler could explain them in a rigorous, cogent manner. Thus, he communicated with the rest of the group from England as they prepared their joint paper.

As Geoff Burbidge recalled: "Margaret Burbidge and I wrote the first draft of B²FH. We deliberately incorporated extensive observations and experimental data supporting the theory, and Hoyle and Fowler worked extensively on the early draft to see that all of the work was covered. There was no leader in the group. We all made substantial contributions, and Hoyle was entirely happy with the result."[10]

On October 1, 1957, their joint paper "Synthesis of the Elements in Stars" was published to great acclaim. It has remained one of the most influential papers in astrophysics.

The B²FH era was a happy time for the Burbidges. Not only did they contribute to a major scientific find, they also reveled in the birth of their daughter, Sarah, who would grow up surrounded by astronomy and cosmology. "It was our life. I was their only child. Most of their friends (were scientists)," Sarah recalled.[11]

Sarah fondly remembers how Hoyle and Fowler were treated as part of the family. "[Hoyle] was fabulous. He was my Uncle Fred. He was very, very funny and very interesting. He paid a lot of attention to everyone. . . . My main recollection of Willy [Fowler] centers on Cambridge, starting around 1968. We would go to Cambridge every summer. All of the socializing centered on Willy, who was a wonderful character. It was a collegial group. Everybody had fun together."[12]

Hoyle's son, Geoffrey, similarly remembers the B²FH team's continued camaraderie, which would last for decades:

> The Burbidges and Willy Fowler were regular visitors to our home, both as colleagues and friends. As they spent long periods in Cambridge, Willy Fowler bought a house there. The Burbidges maybe stayed in college, university or other rented accommodation. As my father preferred to work at home there was a continuous stream of academics through the house. Willy and others, such as Don Clayton (a former PhD student of Willy's), would go walking in Scotland with my father. Having attended some of Willy's parties in Pasadena, I would say their meetings were lively and stimulating. Geoff Burbidge had an acid sense of humour and didn't tolerate fools willingly.[13]

Around the time of the B²FH contribution, the work of the Kellogg researchers examining carbon-12 resonance, headed by the Lauritsens, was featured in the press. Sensing a dramatic story of clashing universe models, on December 30, 1956, the New York Times covered a talk by Thomas Lauritsen at the annual meeting of the American Physical Society about his group's efforts to confirm Hoyle's theory of helium burning and the 7.65 MeV energy level of carbon-12. The method he discussed was bombarding carbon with deuterons to break it up into three helium nuclei. The banner read: "Physicist Makes Helium of Carbon; Transmutation Is Hailed as Helping to Explain Origin of Universe; 'Big Bang' Theory Hit."[14]

That headline, emphasizing connections between the research at Kellogg and steady-state predictions, was somewhat misleading, however, given that stellar nucleosynthesis of heavy elements was perfectly compatible with most versions of the Big Bang. In fact, the ultimate theory of element formation would involve both the Big Bang (for helium creation) and stars (for the higher elements).

DRIVING ABBÉ LEMAÎTRE

By the time the B^2FH paper was complete, Hoyle was understandably exhausted from all the transatlantic travel to and from Pasadena. He realized that he had also taxed his Cambridge colleagues' patience by spending so much time abroad, making use of travel funds, and taking long breaks from teaching his expected roster of courses. In 1956, he took a paid sabbatical from teaching and other university responsibilities, followed by an unpaid leave not too long after that expired. By the end of that year, he had made amends by committing to a heavy teaching load in the spring term of 1957.

Hoyle was thinking seriously about permanently relocating to Southern California, which had become the testing grounds of his bold ideas.[15] Compared to Cambridge, Caltech had far superior resources, such as wider access to computers. It offered its research faculty relatively light teaching loads. The weather would be sunny, and hiking opportunities abundant. Other places in America looked similarly promising. He would be able to escape the staid rituals of Cambridge life and, above all, its politics, which he detested. A particular problem was volatile radio astronomer Martin Ryle, a prolific researcher and opponent of steady-state, who Hoyle felt was out to get him. Ryle headed the radio astronomy unit of Cavendish Laboratory. Tommy Gold, who had been a target of Ryle's venom, had already successfully made such a move to the States, riding a westerly current to Harvard (he later migrated to Cornell for a permanent position).

Frustratingly, though, Hoyle's wife, Barbara, had no interest in relocating to the United States. Fred's powers of persuasion proved of little use. Therefore, at least for the time being, he confined his wanderlust to navigating the mazes of streets and courtyards that abutted the River Cam. Travel was hardly on his mind.

Nevertheless, when invited as the sole British representative to a May 1957 astronomy conference at the Vatican, where other eminent cosmologists would be gathering, Hoyle felt honored and obliged to accept. It meant taking leave, once again, from Cambridge—and irritating some of his colleagues, perhaps—but how could he pass up such an opportunity? The invitation permitted him to bring Barbara, too, so the two took a working holiday. Behind the wheel of their bulky Humber Hawk, the Hoyles beamed with excitement as they made their way across the channel, through the bucolic French countryside, including the beautiful town of Chartres, with its famous cathedral, down the Italian coast, past Pisa's nonorthogonal edifice, and toward the Holy See.

Hoyle likely wasn't fully aware of the politics behind his invitation. Father Daniel O' Connell, director of the Vatican Observatory and one of the conference's key organizers, knew of Pope Pius VII's support of the Big Bang theory, which His Holiness had announced at a 1951 meeting of the Pontifical Academy and a 1952 meeting of the International Astronomical Union. Nevertheless, O'Connell believed that Hoyle, with his solid reputation as a leading expert on stellar dynamics, would be the ideal person to summarize the meeting objectively for the conference proceedings. There were only two potential issues: Hoyle's opposition to the Big Bang and his skepticism about organized religion (as stated, for instance, in his book *The Nature of the Universe*). Fortunately, the pope personally intervened and informed O'Connell privately that Hoyle could attend: "Communicated by voice to Father O'Connell *Ex Audientia* with His Holiness 27 February

1957. . . . Considered what Father O'Connell affirms, Professor Hoyle can be invited."[16]

The conference attracted many notables, including Baade, Fowler, Sandage, Salpeter, Martin Schwarzschild, and even the venerated father of the Big Bang, Georges Lemaître. Hoyle gave several well-received talks, including one about the B²FH theory of stellar nucleosynthesis. The audience, including the aforementioned individuals, asked deep questions about the origin of helium given its great abundance throughout the universe. Schwarzschild suggested that although the B²FH theory was likely correct for the heavy elements, stellar nucleosynthesis couldn't adequately explain all the helium, for which Gamow's theory of Big Bang nucleosynthesis seemed a much better match. Overlooking the abundance issue, Fowler chimed in that he found it amusing that Schwarzschild assigned a well-known stellar process to a "primeval event." To that remark, Lemaître replied, "How do we know hydrogen is the starting point?"[17] Hoyle explained it had to do with hydrogen's overwhelming abundance, which couldn't just be coincidence.

When Barbara and Fred Hoyle found out that Lemaître was heading up to Belgium, which was on their way back to England, they kindly offered him a ride, which he gratefully accepted. The ride was memorable because Lemaître knew the route well and offered many excellent sightseeing suggestions. Remarkably, there was no row about cosmological theories. Rather, the only point of contention had to do with meals: Lemaître insisted that they stop for a long leisurely lunch each day, with ample food and wine. Then he'd get back in the car, fall asleep, and wake up with a bad headache.[18] The Hoyles relented. A more uncomfortable mealtime situation occurred during a Friday dinner at an inn, when Fred made a joke about the Catholic mandate to eat fish — it went down like a lead sinker. When Hoyle saw that Lemaître's fish platter was much larger than his own steak, he remarked, "Now at last I see, Georges, why you are Catholic."[19] Clearly annoyed by the lighthearted jab at a sacred tradition, the priest glared back.

In the same year as the conference and the road trip, Hoyle published his first science fiction novel, *The Black Cloud*. Its positive reception encouraged him to pursue that genre as a second career. His novels not only brought in some extra income but also gave him a way to express his speculative ideas to the general public.[20] For example, in *The Black Cloud*, he included a sly dig at Big Bang theorists. The Cloud (an alien entity) tells the astronomers communicating with it: "I would not agree that there ever was a first member."

The fictional astronomers' response is telling: "They exchanged a glance as if to say, 'Oh-ho, there we go. That's one in the eye for the exploding universe boys.'"[21]

In Hoyle's second novel, *Ossian's Ride*, published in 1959, the main character is an anti-establishment scientist turned spy—and an expert long-distance walker, to boot—who is investigating a mysterious technological transformation that has taken place in an Ireland of the future. One cannot help but note that the protagonist bears some resemblance to Hoyle's own self-image as a clever rebel, wandering through the world resisting authority and seeking truths.

Hoyle later collaborated with his son in science fiction writing. As Geoffrey recalled: "When he got a break in his work, my father often wrote up a draft of some idea triggered by the work he had been doing. For him it was a form of relaxation and escape from the pressures of academic life. Alternatively, it was a means of communicating ideas rejected by the academic establishment. The science was the most important element of our writing."[22]

While Hoyle's career was riding high, Gamow's research program was floundering. For Gamow, the highlight of that period (which, as mentioned, saw him divorce Rho, leave GWU, take up a position at UC Boulder, and get remarried to Barbara Perkins) had to do with his popular science writing rather than with his physics. In 1956, he was honored with the Kalinga Prize for the Popularization of Science, a prestigious award administered by the United Nations Educational, Scientific, and Cultural Organization (UNESCO). Even in that area, Gamow's rival would eventually catch up. Hoyle received the Kalinga Prize in 1967, honoring his own science popularization.

Gamow hadn't said much more about the Big Bang since his work of the late 1940s and early 1950s. Except for its lack of a reasonable prediction for helium creation, the B²FH theory seemed to explain everything he had tried to resolve. He agreed with its premise and tried, unsuccessfully, to think of some aspect to which he could make a contribution. No doubt, that was disappointing.

Moreover, Gamow's habit of drinking to unwind or socialize had turned into a much more frequent state of inebriation fueled by a steady succession of martinis and other alcoholic beverages. In any tight-knit community, gossip spreads. Gamow was already known in the world of physics as being fun-loving, zany, and a practical joker. Sadly, by the late 1950s and the early 1960s, the epithet "drunk" would come up too. For example, astronomer Vera Rubin, who worked with Gamow at GWU and later made groundbreaking discoveries related to the study of dark matter, mentioned to others that she was sometimes embarrassed by his behavior when he imbibed.[23]

When asked in an oral history interview if he thought drinking affected Gamow's reputation, Alpher responded:

Boy that is hard to say. I suspect it did. But I'm not sure but that he wasn't this kind of an outgoing, enthusiastic character even cold sober. And I think that . . . and his great love for physics and the idea that it was all fun, I don't think went over too well with a lot of people who were rather serious about what they did. And I think people misunderstood him; they took his sense of fun and humor, and so on, and turned that into "Well, he can't be that good if he's not that serious about what he's doing, or doesn't seem to be that serious."[24]

In 1958, Gamow planned a trip to Europe with his son, Igor, with the hope of attending the Eleventh Solvay Meeting in Brussels. The theme of the meeting was right up his alley: "The Structure and the

Evolution of the Universe." Hoyle, Bondi, and Gold would be there. They all would have had a chance to discuss ideas. Alas, the encounter was not to be. Gamow was not invited. He attributed his being overlooked to his dismissal of the steady-state in his writings.

Nonetheless, he and Igor had a fun time. As Igor recalled: "In 1958 he took me to Europe, which was a wonderful trip. He'd go through a huge amount of drinks. When he was over-drinking, he was pretty much the same as when he was not drinking. Father was such a Russian. In the end, it really got a hold of him. He was crashing cars."[25]

THE RELENTLESS ROW WITH RYLE

Though Hoyle had won the great respect of British cosmologists and astronomers, not everything was harmonious. Within the competitive world of Cambridge academic culture, he had been forced to deal with Martin Ryle as an increasingly powerful opponent. From Hoyle's perspective, Ryle's mission seemed to be making his life absolutely miserable.

Ryle was a force of nature—for better or worse—a torrent of emotions that might generate incredibly useful energy or leave massive destruction in its wake. The sharp contradictions in his personality baffled even those who admired him. Although he prided himself on his antiwar and antinuclear stance, he had a temperament so violent that if he disagreed with someone during class or a seminar, he was known to hurl objects such as blackboard erasers at them.[26] A frightening example was an incident in which one of his staff members was presenting an analysis of collected data and included certain readings that Ryle thought were extraneous and should have been left out. Mad with rage, Ryle grabbed a marble inkwell and threw it at the man, knocking him out.[27]

Ryle was also antifascist, with a love for democracy that, despite his overall hatred for war, had motivated his heroic efforts during World War II to work diligently at the Telecommunications Research Establishment on radar countermeasures to protect British pilots.[28] Yet, after the war, his management style at the lab with his group had a distinctly authoritarian flavor. He zealously protected and nurtured members of his research team, but he was often hostile to outsiders. Another irony was that his postwar career at Cambridge benefited from the spoils of victory: to set up a radio astrophysics program at Cavendish Lab, he procured radar equipment captured from the Germans.

In addition to Ryle's temperament, his social status may have played a role in his ongoing feud with Hoyle. Whereas Hoyle came from a working-class region of West Yorkshire known for its commerce and industry, Ryle hailed from Brighton, in the south, a seaside center of leisure. Ryle's family was soundly upper middle class; his father, a well-known physician. Hoyle, who worked very hard to rise to his position at Cambridge, disdained the idea of anyone getting ahead even partly because of background or privilege.

According to Hoyle, the row between the two of them began in 1951 when, during Tommy Gold's talk about "radio stars," Ryle reportedly attacked Gold and his ideas in a condescending manner, knocking theorists in general for failing to see the real picture. *Radio stars* was the name astronomers had given to mysterious sources of radio waves that act like transmission towers in space. As was the case with nebula in the time of the Great Debate a generation earlier, the astronomy community was torn over how to interpret these radio sources. Were they objects within the Milky Way or extragalactic bodies? Gold, in his lecture, pointed out the fact that they seemed to be distributed isotropically, or roughly the same in all directions, which made him think that they were all extragalactic. One of the radio sources he spoke about was called Cygnus A, among the first to be discovered.

Ryle, by that time, had constructed a revolutionary new device called an astronomical radio interferometer that could pinpoint the location of energy sources with much greater precision than ever before. He used it to survey a number of radio sources and was convinced that at least some of them, if not all, were inside the Milky Way. Hence his skepticism about Gold's conclusion.

But, in 1952, at the famous International Astronomical Union meeting in Rome, when the pope endorsed the Big Bang, Baade had announced that at Mount Wilson Observatory he had visually identified the location of the Cygnus A radio source. Based on its measured distance of around five hundred million light-years from Earth (since revised to about 760 million light-years; one light-year, the distance light travels in a year, is about 6 trillion miles), it was certainly an independent galaxy, not a Milky Way object.

After Baade's announcement, Hoyle had reminded Ryle that Gold had been right. But actually the truth turned out to be complicated: some radio sources do lie inside the Milky Way, as Ryle had found; others are remote, active galaxies, as Baade and others had found. For some reason, Hoyle believed that Ryle had been "humiliated" by Baade's discovery confirming at least part of Gold's conclusion and then spent the next ten years seeking revenge on Hoyle for bringing it to his attention by counting radio sources in an effort to disprove the steady-state theory.[29]

After Baade's finding about Cygnus A, Ryle did indeed make a concerted effort to map out the number of extragalactic radio sources at various distances. It is hard to believe that he conducted such a grueling mission simply for spite. However, it was true that he had become one of the few voices at Cambridge to cast doubt on the steady-state theory, which, thanks in part to the Astronomer Royal's support and Hoyle's popular radio programs and books, had become the pride of Britain. Ryle's increasingly rigorous studies of the distribution of active galaxies emitting radio signals led him to conclude that the universe was aging, far from remaining in a constant state. (Today we know that many galaxies have active cores during an early stage of their development, fueled by central, supermassive black holes.)

On the basis of a set of data he had collected over several years, Ryle reported his stunning conclusion during the 1955 Halley Lecture delivered at Oxford: "This is a most remarkable and important result, but if we accept the conclusion that most of the radio stars are external to the

Galaxy, and this conclusion seems hard to avoid, then there seems no way in which the observations can be explained in terms of a Steady-State theory."[30]

Steady-state was so popular then in the UK that when Hoyle cast doubt on Ryle's results, because of sparsity of data and poor statistics, many took Hoyle's side.[31] In September 1956, he was able to explain his perspective in a special issue of *Scientific American* called "The Universe." That issue proved to be a print debate between Gamow (who naturally sympathized with Ryle's view) and Hoyle, with back-to-back articles titled "The Evolutionary Universe" and "The Steady-State Universe," respectively. That was the closest to a direct public confrontation Gamow and Hoyle would ever have. The following year, when the monumental B²FH paper appeared, Hoyle was lauded, which seemed to temporarily shield steady-state from further criticism.

The dam broke, however, in February 1961 when Ryle, after collecting much more substantial radio source data reflecting a wide range of distances, delivered a well-publicized talk to which Hoyle and his wife were invited. Hoyle hoped that his invitation implied that Ryle would offer some concession to the steady-state model. Instead, Ryle pronounced that his enhanced data made it absolutely certain that steady-state was wrong. Because reporters were in the audience, primed to gather Hoyle's reaction to the bad news for his model, Hoyle was enraged by what he felt was a blatant setup.[32]

Indeed, newspapers around the world reported Ryle's results as potentially settling the battle between Gamow's "evolutionary universe," or Big Bang theory, and the steady-state model of Hoyle, Bondi, and Gold, in favor of the former. Steady-state supporters were forced to refute the data. In March, Gold, who by then had relocated to Cornell, and Dennis Sciama, who was visiting there, gave talks in which they disputed Ryle's claim that he could accurately pinpoint the distances to the radio sources, thus defending the steady-state interpretation.

Meanwhile, another prediction of the steady-state theory—that matter and antimatter would be produced in equal quantities—was cast into doubt with the launch of the Explorer XI satellite. Its mission was to look for the gamma rays that would indicate the annihilation of such particle–antiparticle pairs. But the satellite found little evidence of such activity. Astronomer Robert Jastrow, in a speech delivered at the National Rocket Club in December 1961, indicated that the gamma radiation level was so low as to "rule out one version of the steady-state of cosmology which holds that matter and antimatter are being created simultaneously."[33]

By the early 1960s, the standard steady-state model seemed on increasingly shaky ground owing to Ryle's radio source distribution data (if one believed in its accuracy) and the new satellite findings, whereas Hoyle's other major project, mapping out the origin of the elements, seemed in excellent shape, with one exception. Neither he nor any of his colleagues working on stellar nucleosynthesis had a solid idea about how the enormous amount of helium in the cosmos was created. The best-case scenario for stars, he found, was that they could create about 2 percent of the helium in the universe, whereas the reality is that the helium content of the universe, as a percentage of all the chemical elements, lies close to 25 percent. He toyed with the notion of interstellar radiation breaking down higher elements into helium but found that such processes were extremely unlikely to produce the enormous amount of helium found in the cosmos.

Finally, while preparing a course in cosmology during the 1963–1964 academic year, Hoyle took a step that would have been virtually unthinkable for him years earlier. He conceded—at least for the time being—that Big Bang nucleosynthesis of helium was well worth exploring. Maybe a "small residue of Gamow's idea," as he put it, was correct after all. To that end, he teamed up with colleague Roger Tayler and produced a paper that updated Gamow, Alpher, and Herman's results for helium (and deuterium) only. They included several modifications based on recent discoveries in particle physics, such as the theory that there are several types of neutrinos.

Hoyle's collaboration with Tayler proved an inflection point in the long-standing drama of Big Bang versus star-driven element creation. His steady-state fortress was starting to crumble, with many researchers, even in Britain, flocking to a universe with a finite age. Given the success of Big Bang nucleosynthesis to account for the helium in the universe—and a fraction of the lithium, as it turned out—and stellar nucleosynthesis for all of the elements beyond that, the consensus leaned toward a hybrid model.

The only thing missing was the smoking gun proving that the Big Bang had actually happened. That would appear soon, with Penzias and Wilson's incredible, accidental discovery of the telltale background radiation that fills all of space, leftover and cooled down from a hot primordial era.

CHAPTER SEVEN

Triumph of the Big Bang

> Once we realized we were seeing noise from the sky nothing else mattered.
>
> —Arno Penzias

THE MOST SERIOUS CHALLENGES TO STEADY-STATE WOULD come in the wholly unexpected discovery by Arno Penzias and Bob Wilson of a radio hiss from the past. They recorded the hiss from a horn antenna and revealed a source of radiation that could not be traced to anything local or immediate. Its temperature profile, of around 3 degrees above absolute zero, was consistent no matter which way they pointed the receiver. Once they ruled out all the noise-generating effects they could think of, including the possibility that pigeon droppings were creating static, they were at a loss.

The man who would finally decode the signal and reveal its primordial origin was Princeton physicist Bob Dicke. Dicke was brilliant at thinking up clever ways to test unusual hypotheses. His research uniquely bridged the gap between theorists and experimentalists in the subject of general relativity and related areas because he kept one foot in each arena. At the time Penzias and Wilson were seeking answers about their signal, Dicke, along with his students David Wilkinson and Peter Roll, was planning his own radiometer to search for evidence to boost a theory he had developed about a hot early universe. He was therefore astonished when he received a phone call from Penzias about the unusual background hiss.

Dicke and his research associate (and former student) P. James E. "Jim" Peebles demonstrated that the background radiation bore the distinctive fingerprints that matched with models showing the cosmos as a once-blazing fireball that has since cooled down substantially because of its expansion. In formulating that theory, Peebles, who did the calculations, "reinvented the wheel," as he has often put it, in reproducing the essence of earlier calculations by Alpher and Herman, which they had conducted when they were exploring the ramifications of Gamow's Big Bang nucleosynthesis idea. Gamow and especially Alpher and Herman would often remind Dicke and Peebles of that fact when the latter's work in establishing proof of the Big Bang became a major news story in 1965.

It was a low point for steady-state, for sure, because even Hoyle was plagued with doubts and had conceded that aspects of the Big Bang might be right after all—before veering back to a modified version of the steady-state and a little-accepted alternative explanation of the radiation that involved tiny shards of iron or graphite (a form of carbon) in space that thermalized starlight by absorbing it and reemitting it as microwaves.

THE FIRE OF THE PHOENIX

Ironically, though, before the discovery of the cosmic microwave background radiation, Dicke was far from a Big Bang acolyte. Rather, he believed that general relativity itself was incomplete. In his view, it didn't adequately explain inertia, the tendency for objects to remain at con-

Princeton experimental physicist Robert Dicke, who was an expert at testing some of the predictions of standard and alternative forms of general relativity and cosmology. CREDIT: AIP Emilio Segrè Visual Archives, Physics Today Collection.

stant velocities along straight lines unless jolted by forces. Dicke thought Mach's principle, the idea that the combined tugs of distant massive bodies produce the inertia effect, needed to be integrated into general relativity. Recall that Einstein had thought so, too, but had never quite achieved that goal.

Along with young researcher Carl Brans, Dicke developed a way of incorporating Mach's principle into a modified form of general relativity. Instead of a gravitational constant that depicted the strength of gravity (for a given distance from a certain amount of mass) as the same for all times and places, Brans and Dicke proposed making gravitational strength a dynamic quantity that would potentially vary throughout time and space. Brans described the model in his PhD dissertation, "Mach's Principle and a Varying Gravitational Constant." Brans and Dicke realized that the expression "varying constant" was an oxymoron but found that it succinctly expressed their goal of loosening an otherwise rigid value. Because the German physicist Pascual Jordan had previously developed a similar model of variable gravity, their combined proposal has come to be called the Jordan-Brans-Dicke model.

Dicke was frustrated at times by the lack of mainstream acceptance of his theory. Most researchers had flocked to the standard solutions of Einstein's equations rather than modified versions. As Dicke wrote to Peter Franke: "I suspect that there are thousands of theorists who would rather believe that the moon was made of cheese than believe that there could be a scalar component in the gravitational attraction."[1] ("Scalar component" refers to a special type of energy field that is variable from point to point, something like a weather forecaster's map of temperatures in a region.)

But even before Jordan, Brans, and Dicke, perhaps the most prominent supporter of a variable gravitational parameter was Paul Dirac, who made his case back in 1937 with the large numbers hypothesis. In a paper, Dirac makes the assertion that various combinations of fundamental constants produce enormous fixed values. He believed that understanding such large numbers was essential for understanding nature. One key prediction of Dirac's theory is that gravity's strength — tied to other parameters rather than being independent, as it is conceived in conventional relativity — has decreased with time. Such diminution leads to bizarre consequences, such as the gradual expansion of Earth throughout the eons as gravity's hold on it relaxes. No one has observed such growth, but in the mid-twentieth century less was known about geology and this idea was not beyond the realm of possibility. Dirac's hypothesis exerted a profound influence on both Hoyle and Gamow, who would each entertain the notion of a variable gravitational constant at some point.

Hoyle's Machian foray began in the early 1960s, in collaboration with Jayant Narlikar, his brilliant PhD student from India. Reviving the Wheeler-Feynman absorber theory of electromagnetism (involving direct action at a distance between particles), they proposed that the mass of any object in space, rather than being a local quantity, depended on distant interactions with all other massive objects in the cosmos. Each massive particle is like a maypole with strands attached to every other massive body. Without those strands, its mass would vanish. In line with Hoyle's penchant for getting the word out about his idea, the Hoyle-Narlikar theory received considerable press coverage:

> What would happen to the solar system if half of the universe disappeared? From Newton to Einstein, most experts have agreed that nothing much would happen except that the sky would have fewer stars. But now British Cosmologist Fred Hoyle says that the sun would shine 100 times brighter and burn the earth to a crisp.
>
> Hoyle is a respected scientist, one of the originators of the theory of continuous creation, which holds that the universe is still being formed by particles that appear out of nothing in empty space. When he presented his new gravitation theory to a packed meeting of Britain's venerable Royal Society, he modestly described his work, done in collaboration with Indian Mathematician Jayant V. Narlikar, as a slight extension of Einstein's theory of general relativity. "We are clearly aware," he explained, "that in putting forward still another idea we may be like small boys trying to steal apples."
>
> Far from a slight extension of Einstein's work, Hoyle's apple stealing is more ambitious larceny. His new theory stems from the Mach Principle, that the mass of every object in the universe is affected by its interaction with every other object. Einstein tried to incorporate the Mach principle in his own scheme of the universe and admittedly failed. Hoyle claims to have succeeded.[2]

The Hoyle-Narlikar proposal was met with criticism from an unlikely source. When Hoyle spoke on the subject at the Royal Society in 1964, Stephen Hawking, then a twenty-two-year-old student of Sciama at Cambridge (who had wanted to work with Hoyle but who was turned down by him), brought up fatal flaws of the theory during the question time following the lecture. According to some reports, Hoyle was rattled by his comments. Hawking later published a paper refuting the theory. Nonetheless, Hoyle was magnanimous in advocating that his department continue to fund Hawking as a student, and later hire him as a professor. Hoyle generally welcomed debates, as long as both parties remained respectful and open-minded.

Despite a mutual interest in Mach, Dicke largely ignored the Hoyle-Narlikar model. As an innovative experimentalist with renegade notions, he was far more interested in testing his own ideas, through creative methods, than in exploring the nuances of competing theories.

When Einstein had attempted a cosmology based on Mach's principle that accounted for inertia guided by distant massive objects, such as stars and galaxies, recall that he selected a closed, isotropic geometry (versus open or flat). That way the universe is finite and bounded, sprawled over the surface of a hypersphere (the four-dimensional analogue of a sphere). Consequently, the combined pull of all of the objects in space produces a finite effect, which might be equated with inertia. It is just enough to steer frictionless bodies, such as hockey pucks on absolutely slippery ice, to slide along perfectly straight lines—but not the infinite effect that an endless universe (open or flat) would produce.

Similarly, in his own efforts to bring Mach's principle to fruition, and thus fulfill Einstein's unrealized dream, Dicke similarly advocated a closed geometry. As Friedmann and Lemaître each demonstrated, and Robertson cataloged, in the absence of a cosmological constant, general relativity predicts that closed, isotropic universes must begin at a mathematical point, expand to a maximum girth, and then contract down to a point, like a round balloon inflated and then deflated. If the entirety of the universe's mass was focused at a single mathematical point, that would represent a monstrosity of infinite density, called a singularity.

Philosophically, Dicke hoped that the initial and final singularities could be avoided. There was no reason to presume that a Jordan-Brans-Dicke model, with variable gravitational strength, would necessarily begin and end at such infinitely dense points. With the luxury of dialing up and down the clustering ability of gravity, such models could be tweaked to avoid singularities. Even in the case of standard general relativity, where gravity's strength is set in stone, Dicke shared with many other researchers of his time the belief that realistic physical models (with irregularities that make them not completely smooth) would lack singularities. According to such reasoning, just as deflating a basketball would not lead to a perfect point but rather to an irregular glob, the nascent and final moments of the universe would similarly be blob-like rather than point-like.

In that case, Dicke thought, perhaps the ultimate crushing moment of the current cosmic era wouldn't be so final. Rather, the universe would bounce back, in a new age of growth—like the mythical phoenix rising from the ashes to live for another cycle. Dicke found such persistence to be very satisfying, because it avoided the need for a beginning and end of time. In that manner, it was similar to steady-state. Along the lines of a similar model by Richard Tolman, Dicke deemed his approach the "oscillating universe."

As Dicke wrote to Arthur Moor, who inquired about the reasoning behind the oscillating universe: "It appears that there are religious, philosophical, biological, geological, and mathematical aspects to the idea of an oscillating universe in addition to the physical and astronomical. On the religious side one could say that the concept of a continuously regenerating periodic universe is almost Hindu in its conception."[3]

One prediction of Dicke's oscillating model is that the universe would build up entropy (a measure of disorderly waste energy) at the end of each cycle, like a heavily used dryer accumulating more and more lint. Dicke expected, therefore, that the universe at the beginning of our current cycle was full of cold thermal energy leftover from its previous incarnation. Perhaps, he wondered, such a tepid bath of radiation has lingered in space and might possibly be detected. As Peebles explained: "[Dicke] had an idea about the early universe that would have made it hot. What was the universe doing before it expanded?

A bounce would be highly dissipative and irreversible, and would produce a lot of entropy in the form of thermal radiation. It would be a consequence of the radiation present in the previous cycle and its behavior during the bounce. I was charged with the theoretical consequence of finding the radiation."[4]

Dicke would later write to Hawking, who, along with George Ellis, was engaged at Cambridge in a theoretical study of the conditions under which the universe would experience an initial singularity: "Our work on cosmic thermal radiation was motivated by the guess that the singularity could be avoided in the closed space."[5]

Dicke was apparently unaware, at the time, of the work by Alpher and Herman predicting that space was filled with a bath of cooled-down relic radiation leftover from a hot primordial era. He had indeed been in the audience for a talk at Princeton by Gamow about the early universe but had mistakenly thought Gamow had said that the Big Bang was cold, not hot. "I'd heard Gamow give a colloquium talk here, and I think it was either a colloquium or possibly a Sigma Xi talk or something of that kind, in which he described his ideas about heavy element formation in the early universe. . . . Whether this was a preliminary view or what, I distinctly remember him describing this as a completely neutron-filled universe and starting out cold, so the idea of it being hot hadn't [been mentioned]. . . . I didn't realize he had that idea. We should have taken this up, but just didn't."[6]

It is unlikely that Gamow would have ever suggested that the early universe was cold, given that it needed to be blazing hot for element formation. Therefore, Dicke must have misinterpreted something Gamow said. Or perhaps he had been thinking about the Mayer-Teller "polyneutron" model instead, which starts out cold. As a consequence, in developing his own theory of cosmic radiation, it didn't occur to him to look at the papers by Gamow or his associates Alpher and Herman.

Dicke's pursuit of evidence for cold radiation drove him to assign his students Wilkinson and Roll the task of constructing a radiometer able to detect such signals. But before that was built, he received the unexpected request from Penzias to take a look at the graphs of the mysterious noise he and Wilson had detected using the twenty-foot Bell Labs horn antenna in nearby Holmdel, New Jersey.

HISS OF THE COSMOS

The accidental cosmic discovery was far from the only miracle in Penzias's life. His opportunity to live safely and peacefully in the United States collecting radio data was, for him, miraculous enough, given his tumultuous childhood. Born in Munich, Germany, shortly after Hitler ascended to power in 1933, he and his family, as Jews, were subject to brutal oppression and hatred. Shortly after he turned six, his parents informed him that he and his younger brother were eligible for Kindertransport, a program enabling Jewish children to leave Germany and stay temporarily in England with the hope that their parents would soon join them. In the case of the Penzias family, their plan was to use England as a stopover on the way to emigrating to the United States. As it turned out, the stopover was for six months, during which time young Arno began to learn English. Luckily, the family would indeed reunite and take a ship to New York, where they'd resettle. As Penzias recalled the ocean voyage: "I was in this boat and we were in cabins. This was the most horrible experience. In the middle of the night I was woken up and told that my brother was crying. I was 6 years old. I had to calm him down."[7]

Arno's family encouraged him to pursue a career in science. He took advantage of the free tuition at City College of New York before earning a PhD at Columbia University under the supervision of Charles Townes (who would later win a Nobel Prize for inventing the maser, the predecessor to the laser). The topic of his thesis was radio astronomy.

After receiving his doctorate in 1961, Penzias found a position at the Holmdel location of Bell Labs, which was in the midst of Project Echo, an experiment in long-distance radio and telephone communication. By bouncing signals off the Echo 1 satellite, launched in May

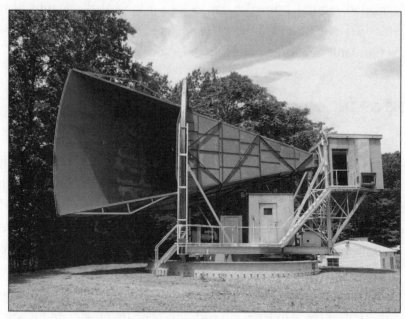

The horn antenna at Bell Laboratories in Holmdel, New Jersey, with which the cosmic microwave background radiation was accidentally discovered in 1964 by Arno Penzias and Robert W. Wilson. CREDIT: Photograph by Paul Halpern.

1960, the first satellite phone call was transmitted between NASA's Jet Propulsion Laboratory in Pasadena and the horn antenna, specially constructed for the project, equipped with a microwave receiver to pick up the signal. Once Echo 1 was retired and the project moved onto a new phase, the antenna became available for use in astrophysics. Penzias, along with his new associate Bob Wilson, jumped at the chance to use it to look for a hypothesized radio source halo around the Milky Way.

Wilson, who was born in Houston, Texas, and three years younger than Penzias, arrived at Bell Labs after earning a bachelor's degree from Rice University and a PhD in physics from Caltech. Ironically, although he would help prove the Big Bang and rule out the steady-state, his original inclination was to support the latter. "I didn't like the idea of a beginning or end of the universe," recalled Wilson. "I liked the idea of the universe going on forever."[8]

Arno Penzias, co-discoverer of the cosmic microwave background radiation. CREDIT: AIP Emilio Segrè Visual Archives, Physics Today Collection.

Wilson was at Caltech during a time when Hoyle was a visiting faculty member. "I had a course from Fred Hoyle in cosmology," he noted. "It wasn't something I was unusually interested in, but I wanted to know about it. I don't remember Fred pushing steady-state so much, but that was his thing. Most of the people at Caltech seemed to support Big Bang."[9]

Wilson also came into contact with other members of Hoyle's circle. Willy Fowler taught Wilson's eight a.m. course in nuclear physics, quickly filling the blackboard each session. John Bolton, a British-Australian radio astronomer, was Wilson's research supervisor. Bolton had discovered some of the first radio stars. When he was refused entry to Cavendish during a visit to Cambridge, reportedly because of Ryle's paranoia about outsiders, he met—and became lifelong friends with—Hoyle.

Wilson's thesis project was to map radio sources in part of the Milky Way using a receiver and a Dicke switch (a noise-reducing system invented by Dicke). Thus, his expertise proved a great match for Penzias's when, in 1963, they started working together in Holmdel. As Wilson recalled, "He tended to think of the big picture. I tended to think of detailed things. I started working on the rest of the electronics to make it stable for radio astronomy. I set up the switching system."[10]

Sometime later, when Penzias was flying to a meeting in Canada, he happened to strike up a conversation with Bernie Burke, a researcher at the Carnegie Institution in Washington. When Penzias mentioned his project to map the periphery of the Milky Way, Burke was dubious. "There's no halo around the galaxy," he said. "You are wasting your time."[11]

Penzias would later admit that Burke was right. "Originally I thought there must be something around the Milky Way," said Penzias. "It didn't materialize."[12]

While they were looking for the nonexistent halo, Penzias and Wilson noticed a constant radio static background that persisted no matter which way they aimed the antenna. After ruling out possible environment sources such as radio noise from New York City, radiation that had persisted from hydrogen bomb testing in the Pacific, and defects in the antenna itself, they noticed that several pigeons had roosted within the instrument and were dirtying the space with droppings. So, they ran out and bought a metal pigeon cage, caught the avian perpetrators, and cleaned up the mess. To their disappointment, the hiss was still there.[13]

At one point Burke asked Penzias, "What's going on with your crazy project?"[14]

When Penzias told him about the hiss, Burke recounted that his postdoc Ken Turner had just attended a talk by Jim Peebles at the Applied Physics Laboratory in which Dicke's scheme to detect background radiation in the universe was mentioned. Turner had been a member of Dicke's group and, a good friend of Peebles, remained interested in their projects. As Turner recalled:

Jim and I were both graduate students at Princeton, and members of Bob Dicke's Gravity Group. (Dicke was major professor for both of us.) As I recall, we both entered and got our PhDs the same year,

1957 and 1962, respectively. Jim and his wife, Alison, and I and my wife, Gretchen, were all within a year or so the same age, and were good friends and in frequent social contact.

I heard that Jim was giving a talk at the Johns Hopkins Lab up the road toward Baltimore, and of course I went. I hadn't been in contact with Jim for some years and was eager to see what he and the Dicke group were up to. Jim always gave a good talk and this was no exception. I was excited by Jim's talk and told Bernie all about it.[15]

Once Penzias found out about Dicke's work, he was excited that the hiss might be of astronomical significance. He picked up the phone and called Dicke at his Princeton office.

SCOOPED!

When Penzias rang, Dicke was enjoying a brown-bag lunch meeting in his office with Peebles, Roll, and Wilkinson. They listened intently to Dicke speak and quickly realized something was up. "We were hearing Bob on his end of the line," recalled Peebles. "When Bob got off the phone he told the group, 'We've been scooped.'"[16]

Soon, Dicke, Wilkinson, and Roll drove up to Holmdel to inspect the scene of the discovery. They wanted to make sure that all other possible culprits for the noise had been eliminated. By explaining everything they had done to rule out other effects, Penzias and Wilson addressed Dicke's questions to his satisfaction. As Wilson remembered the visit: "They went up and looked around. They were convinced. After explaining the waveguide things, we came down to the conference room and Bob Dicke gave a talk."[17]

Dicke's talk was not about the standard Big Bang but rather about his oscillating universe idea. Penzias and Wilson listened intently to his explanation, which was centered on the idea "that after multiple Big Bangs there would be entropy piling up."[18]

Alison Peebles, Robert W. Wilson, and P. James E. Peebles at a 2018 Princeton University celebration of the legacy of Robert H. Dicke. CREDIT: Photograph by Paul Halpern.

"I was in awe of Dicke," stated Wilson.[19] Nonetheless, the idea that cosmology—which he had considered rather abstract compared to radio astrophysics—actually made a testable prediction seemed unreal.

When Dicke, Wilkinson, and Roll returned to Princeton, recalled Peebles, "they came back convinced that Penzias and Wilson had done the best job possible ruling out that the hiss came from anything local."[20]

Wilkinson and Roll continued to work on a Princeton radiometer, with the altered mission of independently verifying Penzias and Wilson's results. Meanwhile, Dicke assigned Peebles the task of interpreting the results within a theoretical framework. Peebles was the perfect theorist for the job.

Peebles hails from St. Boniface, a suburb of Winnipeg, Canada, where he was born in 1935. From the youngest age, he wanted to know how things work. As a child, he loved playing with gadgets. "One of my earliest memories," he recalled, "was throwing a tantrum because

I wasn't allowed to put together the coffee percolator. I loved to take things apart and put them back together."[21]

After obtaining an undergraduate degree in physics from the University of Manitoba, as a result of his exemplary performance, Peebles landed in a graduate slot at Princeton. He made the journey of thousands of miles with the intention of becoming a particle physicist and the presumption that it would only be a temporary move. Luckily for the world of relativity and cosmology, he found out from an acquaintance about Dicke's research. That suggestion not only lured him into those fields but also enabled him to walk in on the ground floor of an emerging Princeton enterprise aimed at exploring the universe that has been thriving ever since: "To my intense good fortune, I gave up on particle physics and joined Bob's group."[22]

Turner distinctly remembers meeting Peebles when they were both "Dickie birds" (as members of Dicke's group called themselves). He described Peebles as "tall, thin, Canadian, with a good sense of humor, very bright, and a theoretician by inclination."[23]

Although Peebles would prove to be a remarkably independent thinker, full of exceptional insight, for his graduate work he conformed to Dicke's conjectures of "changing constants" in nature based on the large numbers hypothesis and Mach's principle. The title of his dissertation was "Observational Tests and Theoretical Problems Relating to the Conjecture That the Strength of the Electromagnetic Interaction May Be Variable."

After completing his PhD, Peebles decided to remain at Princeton, first as a postdoctoral researcher in Dicke's group and then as an assistant professor. That's when he educated himself about the possibility of observational tests of more standard cosmologies, namely, isotropic, homogeneous cosmologies with a Friedmann-Lemaître-Robertson-Walker (FLRW) metric (the type that Gamow also had studied) that began as hot fireballs.

The steady-state theory, on the other hand, never interested Peebles. "We didn't discuss it much," he recalled. "There was mild contempt for the steady-state theory. The first time I heard a lecture on the steady-state theory I was skeptical. They just made that up [I thought]. It just sounded silly."[24]

Peebles did have respect for Hoyle's stellar nucleosynthesis work, however. "He was a brilliant physicist," Peebles said. "His work on stellar structure and evolution was transformative."[25]

Even before he learned about Penzias and Wilson's discovery, Peebles was interested in some of the implications of a hot early universe. One idea that occurred to him—he didn't realize it had been explored before—is that the fiery early conditions of the universe offered a perfect furnace for forging helium from hydrogen. Around the time he was working on the project, Hoyle and Tayler's paper "The Mystery of the Cosmic Helium Abundance" appeared in *Nature*. Recall that it was the first work in which Hoyle conceded that a hot Big Bang might provide a solid explanation for the vast amount of helium in the universe. The article referenced two papers on the Big Bang nucleosynthesis work, one the classic 1948 "alpha, beta, gamma" article by Gamow and Alpher, and the other, a 1953 paper by Alpher, James Follin (a researcher who worked with Alpher and Herman on some of their later papers), and Herman. As Peebles recalled: "One day Bob said to me: 'you better go look at this article in *Nature*.' The main theme of their paper is that one can understand the helium as the result of Gamow's hot big bang. It was at this point I learned that I had been reinventing the wheel—that George Gamow had earlier come across the same idea."[26]

Peebles was informed about the early Big Bang nucleosynthesis work in another way, too. Early in 1965, he wrote a draft of a paper on the subject of helium production in a hot fireball and sent in to *Physical Review*. The journal editor, Simon Pasternack, knew much about the history of helium production in the Big Bang, including the work of Gamow, Alpher, Follin, and Herman, and selected an anonymous reviewer familiar with the literature. On the basis of the referee's report that the paper's calculations reproduced some of the earlier results of those researchers, *Physical Review* rejected the draft and apprised Peebles about the chronology.[27]

Even with some knowledge of the early history, Peebles apparently remained unaware of Alpher and Herman's specific prediction that the cosmos would be filled with a cold radiation bath with a temperature of roughly 5 K (degrees above absolute zero). Peebles would later chide himself for "poor homework."[28] Thus, after finding out about Penzias and Wilson's discovery, when Dicke assigned him the task of supplying the theoretical component of a paper on the subject by the Gravity Group, Peebles completed his own calculations from scratch. He compared the approximate 3 K temperature suggested by the profile of the radio hiss with estimates based on two different scenarios: Dicke's oscillatory model and a single, hot Big Bang fireball. By that time, the Hubble constant had been calculated with greater accuracy, allowing him to use a reasonable value of about 10 billion years since the time of the Big Bang. (We now know that the age of the universe is close to 13.8 billion years.) The papers of Dicke, Peebles, Roll, and Wilkinson, with their analysis of the cosmic microwave background radiation (CMBR) and details about the Princeton radiometer under construction, and of Penzias and Wilson, with an account of their horn antenna radio hiss discovery and the specifics of their data, appeared in the May 1965 issue of *Astrophysical Journal*.

Right after the teams submitted their papers, Walter Sullivan, the main science reporter for the *New York Times*, who had written before about cosmology, called Penzias with a host of questions about their discovery. By then, it had been a year since Penzias and Wilson had first detected excess noise with the horn antenna.[29] Wilson knew about the upcoming article but put it out of his mind. But on May 21, when splashed across the front page of the *New York Times* above an image of the Holmdel horn antenna was the astonishing headline "Signals Imply a 'Big Bang' Universe,"[30] the magnitude of their finding finally hit home. As Wilson remembered that eventful morning: "My father came to visit. The next morning he got up, walked down to the pharmacy, bought a *New York Times*, and there was a picture of our antenna on the front page. [At that point we knew that] the world is taking interest in this. I better learn some more cosmology."[31]

Of course, once a story is featured on the front page of the *New York Times*, the news spreads around the world. An ailing Lemaître, who would live only one more year, was reportedly told about the finding.[32] He did not apparently issue any public statements about it, however. With the discovery of strong evidence for the Big Bang, the wild hypothesis he had advanced in his youth had aged into a respectable part of science. From that point forward, its attributes and consequences would be rigorously tested in a manner that had been impossible earlier when he had suggested the notion—including, within decades, with powerful instruments launched into space.

CHAPTER EIGHT

The Point of No Return

> I had this strange feeling of elation and I couldn't quite work out
> why I was feeling like that. So I went through all the things that
> had happened to me during the day—you know, what I had for
> breakfast and goodness knows what—and finally it came to this
> point when I was crossing the street, and I realised that I had a
> certain idea, and this idea [was] the crucial characterisation
> of when a collapse had reached a point of no return, without
> assuming any symmetry or anything like that.
>
> —ROGER PENROSE, Nobel Prize interview

DENNIS SCIAMA WAS A PRAGMATIC SCIENTIST. HE VEHEMENTLY defended the steady-state theory for more than a decade, bolstering Hoyle, Gold, and Bondi as they battled numerous opponents. But then, faced with Penzias and Wilson's extraordinary discovery of a uniform radio hiss consistent with cooled-down radiation from a fiery universal beginning, he suddenly and firmly switched camps. He thereby aligned himself with the mainstream scientific community's conclusion that such unmistakable evidence resolutely supported the notion of a hot early universe and contradicted the steady-state picture of an ageless reality. From that point forward, Sciama joined most of his peers as an ardent supporter of the Big Bang. As he later explained: "It is partly my nature to be a bit passionate, so once I decided I liked the steady-state theory, then even though we didn't know it was true . . . I would get very worked up, particularly if the hostile evidence was rather weak, as at the time I believed it was. Later it mounted up, and we all recanted at different times at our own chosen moments as it were."[1]

Sciama wrote to Dicke with his personal confession: "As you may have heard I have recanted from the steady-state theory, and have taken such a liberal dose of sackcloth and ashes that I am now more orthodox than the orthodox (though I don't suppose this phase will last long). Anyway you can now tell Peebles that I now nearly believe that the excess background has a black body spectrum."[2]

Sciama's conversion came at a critical time in the history of modern cosmology. A key question about the early universe needed to be resolved. Ideal models of the universe—perfectly homogeneous and isotropic like a flawless billiard ball—start off as a singularity, a mathematical point of infinite density. That's because if you take a solid sphere of mass and shrink it down uniformly (without shedding any of its material), its radius gets smaller and smaller while its mass stays the same. Consequently, its density—mass divided by volume—rises to a greater and greater value. At the limit of such contraction, the radius is zero and the density blows up to infinity. Reverse that scenario in time, and one might imagine a perfect ball of mass and energy, embodying the observable universe, emerging from a singularity and expanding until it reached its current scale. But what if the universe wasn't absolutely uniform? In that case, it was thought until the mid-1960s, all bets are off. Perhaps, many scientists believed, any irregularities would disrupt the model enough that there wouldn't have been an initial singularity after all.

It is like someone playing a game of darts and being told not to aim at the bull's eye because if all the darts landed precisely at that point they'd make too great a hole. Realistically, for a typical player, that would never happen. Similarly, what were the chances the entire universe was so regular that—in running its evolution backward in time—all of its material would contract down to the same infinite dent? Seemingly, it was thought, not very likely.

Neither Hoyle nor Gamow liked the idea of singularities in space. Hoyle's universe avoided them by lasting forever. Gamow's model imagined ylem, a simple primordial substance, as the starting point rather than a point of infinite density. Dicke, as mentioned, similarly eschewed singularities. Rather, he favored an oscillating universe, for which radiation persisted from cycle to cycle.

Sciama profoundly influenced two bright researchers, Roger Penrose and Stephen Hawking, to take up fundamental questions in general relativity, including developing general theorems about its workings. Unlike Hawking, Penrose didn't complete his PhD research under Sciama's supervision. Nevertheless, Sciama inspired him to shift his focus from mathematics to mathematical physics. Penrose's appreciation of Sciama was such that he would inscribe his book *The Road to Reality*: "I dedicate my book to the memory of Dennis Sciama who showed me the excitement of physics."[3]

Through Sciama, Dicke learned about the research that Penrose and Hawking had completed regarding unavoidable singularities in general relativity. It was jolting, no doubt, for Dicke to see that their results challenged his oscillatory universe idea. If an unpreventable singularity marked the beginning of time, no previous cycles would be possible.

Penrose's work—for which he would be honored with the 2020 Nobel Prize in Physics (shared with two other researchers)—had to do with highly compact objects that would later be called "black holes." Hawking would make use of Penrose's results by reversing the direction of time, modeling the Big Bang instead. He developed a brilliant proof that the Big Bang, no matter how irregular the universe, must have begun with a singularity.

THE ULTIMATE LIMIT

Black holes represent one of three possible final scenarios for the cores of stars. In the first scenario, a star about the mass of the sun, after it finishes its hydrogen burning, swells into a red giant as its core shrinks. At some point, when all the electrons reach their lowest energy states, a quantum law called the exclusion principle blocks further shrinkage of the core. Once the star's outer envelope has gradually dissipated into space, the core remains stable and becomes a hot but dim white dwarf star.

A second endgame, for stars much more massive than the sun that have a core at least 1.4 times solar mass (called the "Chandrasekhar limit"), involves enormous self-gravitation overcoming the electrons' quantum blockage, leading to a far more rapid and drastic shrinkage of the core. The catastrophic implosion of the core triggers a sudden explosion of the outer envelope as a supernova burst. The electrons of the core jam together with its protons and neutrons to form an ultracondensed uniform state, mainly of neutrons, called a "neutron star."

Penrose's research in 1965 concerned a third scenario (originally explored by Robert Oppenheimer and his student Hartland Snyder in 1939) in which a stellar core is so massive that its self-gravitation is strong enough to overpower all resistance mustered by quantum exclusion rules. The neutrons in a neutron star would be crushed together into an amorphous pulp. For a perfectly symmetric case (like a flawless crystal ball), that collapse would continue indefinitely until the entire mass of the core became concentrated in a single central point of infinite density: a mathematical singularity. But what about more realistic cases of stars that weren't perfectly symmetric? Could they avoid that ghastly fate? As Penrose proved, no matter its shape, as long as the core is massive enough, the collapse proceeds down to a singularity. The star would become a black hole, so named because its gravitation is so strong, not even light can escape its grasp.

Hawking took Penrose's research on singularities and applied it to the early universe. He imagined running the Big Bang backward in time and comparing its reversed expansion to the contraction of a massive stellar core down to a black hole. Curiously, he found that under certain circumstances the two behaved essentially the same. Hawking determined that for a broad range of situations, the Big Bang must have begun as a singularity. That meant that no prior incarnation of the universe could have existed before that point of infinite density and left any traces of its properties. The simplest conclusion, therefore, was that space and time itself began in the Big Bang.

When Sciama presented Dicke with Hawking's proof that the Big Bang was the absolute beginning of everything, Dicke wondered if there might be an exception that would allow for an oscillatory universe. As Hawking's research progressed, including work in collaboration with George Ellis, possible loopholes were ruled out one by one. In the face of an unavoidable initial singularity, Dicke's oscillatory universe no longer seemed viable.

In July 1965, Dicke and Peebles flew to London to attend a conference on general relativity and gravitation at Imperial College. Hoyle was also a participant.[4]

Dicke had met Hoyle at least one time before, at a 1961 summer school in Varenna, Italy, and had nothing but positive things to say about him. He defended Hoyle when a report in *Popular Science*, written by C. P. Gilmore, on Penzias and Wilson's discovery painted him with harsh words: "The first [alternative to Lemaître's model] to gain attention was developed by a brash, tousle-headed, cocky young British mathematician named Fred Hoyle."[5]

In response to the article, Dicke wrote to Gilmore, "I do feel that your characterization of Fred Hoyle is highly accurate [sic]. I consider him to be neither 'brash' nor 'cocky' but rather 'imaginative' and 'quiet.'"[6]

At the London conference, Dicke, Peebles, and Hoyle had a brief, friendly informal discussion about the CMBR.[7] Hoyle showed no sign, at that point, of being dubious of their interpretation. He was still trying to sort out, in his own mind, what it meant. Like Peebles, he was focused on the question of how the hydrogen in the universe was produced. Ordinary stellar processes, he had realized by then, couldn't possibly explain its abundance. By hook or by crook, it must have emerged in a place other than the familiar stars.

THE COAL-BLACK SKY OVER MINERS' COUNTRY

By the time the seminal papers by Penzias and Wilson and Dicke, Peebles, Wilkinson, and Roll appeared, highlighted by the press coverage, public interest in the debate between the Big Bang and steady-state models—and its apparent resolution by the CMBR finding—was arguably at its peak. The drama surrounding whether or not the universe had a true beginning resonated with those fascinated with the deepest questions surrounding life and death. The notion that a glow from the past could unravel the mystery of the origin of everything seemed absolutely incredible.

Dicke received tons of requests to give talks about the microwave background discovery and the Big Bang. However, he saw that facet of his group's research as Peebles's domain theoretically and Roll and Wilkinson's experimentally. Therefore, he generally declined such talks and suggested those researchers instead.

As Peebles recalled: "I presented lots of colloquia and seminars in those days. I don't remember ever being told that Bob had recommended me but he certainly might have. He tended to get a young member of his group started on a project and then let them get to it more or less on their own. A natural continuation of that practice would have been to let them present their results on their own. In my recollection Bob did present colloquia on what he was doing, along with systematic publications of results."[8]

In conveying their results to the public, Dicke's group, largely guided by Peebles, the group's theoretical point man, adopted the media's favored terminology "hot Big Bang" (which, as mentioned, was coined by Hoyle) and "fireball." Lemaître's "primeval atom" and

Gamow's "ylem" (which was never popular outside his circle) largely fell by the wayside. The term *fireball*, which had come to be associated with nuclear bomb tests, offered a misleading vision of the early universe, painting it as a kind of explosion in space rather than as an expansion of space itself. Noting the new meaning, disarmament expert G. J. Ringer wrote to Dicke in July 1965 questioning his use of the term.[9] Dicke asserted that "fireball," in his group's usage, had nothing to do with bombs. Despite recognition that the use of "hot Big Bang" and "fireball" were not perfect, Dicke's group—disinclined, perhaps, to buck popular trends—persisted in using such evocative terminology.

One indication of how much the work of Dicke's team captured the public imagination, even in unexpected places, is a letter he received from George Pothering, a teenager from the town of Pottsville in the coal mining region of Pennsylvania. "Dear Dr. Dickie [*sic*]," the letter read. "I am a junior in high school, and for a science project this year, plan to do a research paper on the 'Big Bang' theory of the universe. I'd like to know your opinion concerning this theory, as well as the probability of it in relation to other theories."[10]

Pothering, now a professor of computer science at the College of Charleston, South Carolina, recalled the background of the letter:

> Since first hearing about the launch of Sputnik, when I was 8 years old, I was interested in everything about the space race, and about stars, planets, and the universe in general. For a science fair project, I decided to do some research into the origins of the universe. Using only what I could find in our local library, which wasn't much since Pottsville was a coal-mining town then with limited access to research literature, I became familiar with the big-bang vs steady-state ideas. After seeing Prof. Dicke's name appearing repeatedly in what I was reading, I decided to write him to see if I could learn more from him. I was doubtful that he would respond to a request from a high school student, and was delighted when he kindly sent me one of his papers.[11]

Disappointingly, despite Dicke's help, young George didn't win first place in the science fair for his Big Bang project. "Alas a narrative presentation could not compete with petri dishes of colorful molds and bacteria."[12]

Whereas *fireball* is hardly used today, the term *Big Bang* continued to be indelibly associated—in scientific discourse as well as in the media—with how the universe began. No one has been able to come up with a better term. Indeed, in 1993, *Sky and Telescope* magazine launched a competition to find a fitter name. Despite 13,099 entries, no superior epithet emerged.

QUASARS AND LITTLE BANGS

Knowing that he couldn't continue to defend the original steady-state theory in the face of the overwhelming support for the Big Bang, Hoyle tried to come up with a solution that preserved the essence of the idea that the cosmos remains roughly similar over time while incorporating new evidence related to galactic radio sources, the helium abundance, and the microwave background. Two ideas emerged that would over the following decades form the basis of a novel construct called the "quasi-steady-state theory." The first was the concept of "little bangs": the release of matter into space by means of explosions in the cores of galaxies or in even larger groupings of galaxies such as clusters and superclusters. The idea was to show how all the helium in the universe could have been produced in local ways, which would maintain the overall sameness of the universe throughout time. The second notion was that, before asserting that the microwave background was a cosmological phenomenon, scientists needed to consider more immediate, less dramatic possibilities. Hoyle eventually settled on the idea that space is full of tiny cosmic needles or whiskers, made of either graphite or iron, that absorb starlight and reemit it in a way that precisely matches the profile for the 3 K temperature Penzias and Wilson had found. By this point, most of the mainstream astronomical

community didn't even follow Hoyle's cosmological research, thinking it so far afield.

As physicist Freeman Dyson, who had some maverick ideas himself,
speculated, "I assume that [Hoyle] continued to disbelieve the evidence
for the Big Bang because he was emotionally committed to the steady-
state cosmology."[13]

Wilson made a similar assessment of Hoyle: "He went to his grave
still thinking he could salvage steady-state."[14]

Hoyle's constant rejoinder to criticism of his speculations was that
he didn't want to follow the herd. As Sarah Burbidge pointed out: "In
[Geoff Burbidge, Narlikar,] and Hoyle's book, there is a photo of a
bunch of geese following a lead goose."[15]

Hoyle's notion of little bangs derives from the prescient speculations
of Soviet Armenian astronomer Viktor Ambartsumian about the centers
of galaxies, introduced at the 1958 Solvay Conference (which Hoyle
attended). Ambartsumian rightly suggested that some galaxies have
"active galactic nuclei," meaning their compact central regions release
enormous quantities of radiation into space. Extraordinarily massive
bodies of unknown identity in those cores, he speculated, cause such
activity. Today we know that supermassive black holes lie at the center
of many galaxies.

The mechanism Ambartsumian proposed for how such massive
central objects in active galactic nuclei released their energy proved
not quite right, however. We now know that supermassive black holes
gobble up nearby material and release enormous quantities of radiation
as these acquired substances plunge into a hole's gravitational well. Ambartsumian speculated that, in contrast, the unknown massive bodies
underwent explosions that hurled matter and energy into space. Astronomers see that energy as radio emissions.

Around the time of Ambartsumian's proposal, astronomers were
accumulating evidence for radio sources of unusually high intensity —
like high-beam headlights among the usual low-beam radio stars.
One of the key researchers in that domain was John Bolton, graduate

supervisor of Wilson and friend of Hoyle. Another was Maarten Schmidt, who made a crucial breakthrough in 1963 using the two-hundred-inch reflecting telescope at Palomar Observatory when he identified a visible object corresponding to one of the intense radio sources. Based on the redshift of its spectral light, he determined that it was likely a large, extremely distant body that, based on Hubble's law of expansion, emitted its radiation roughly two billion years ago. He soon identified more such bodies and dubbed them "quasi-stellar radio sources." Today we call them quasars and identify them with active galactic nuclei from the cosmic past.

Quasars were one of the main topics of discussion at the first Texas Symposium on Relativistic Astrophysics, held in Dallas in December 1963. Because the meeting took place only weeks after the assassination of President John F. Kennedy in the same city, participants understandably were subdued. Nevertheless, the meeting was very successful. In addition to all the talk about quasars, Roy Kerr announced an important solution to Einstein's general relativity theory that described what have become known as rotating black holes. Hoyle, Gold, Fowler, and the Burbidges all attended—with Gold offering a long, warm tribute to Hoyle's contributions to astrophysics during an after-dinner speech.[16]

Young researcher Robert "Bob" Wagoner, who had become a great fan of Hoyle and Gold during his undergraduate days studying electrical engineering at Cornell, was also at the meeting. Wagoner had attended Hoyle's Messenger Lectures at Cornell on cosmology, which influenced him to switch tracks and begin to pursue astrophysics.

As Wagoner remembered the meeting: "I was very fortunate to be able to attend the first Texas Symposium on Relativistic Astrophysics with my PhD thesis advisor, Leonard Schiff. It was held three weeks after and a few blocks from the Kennedy assassination. It was very stimulating to hear the latest ideas about the recent discoveries such as 'quasi-stellar radio sources.' I remember Roy Kerr's talk about his 'algebraically special' solution to the Einstein field equations. Essentially no one there believed that it would be the unique description of a rotating black hole."[17]

In 1964, Gamow, who was also baffled by the discovery of such powerful, distant radio sources, penned a humorous poem about quasars:

> *Twinkle, twinkle quasi-star*
> *Biggest puzzle from afar*
> *How unlike the other ones*
> *Brighter than a billion suns*
> *Twinkle, twinkle, quasi-star*
> *How I wonder what you are.*[18]

In 1965 and 1966, Hoyle began to view active galactic nuclei as a possible way to salvage the steady-state, in a modified form. Envisioning a mechanism similar to Ambartsumian's explosions, he imagined how massive central objects in the center of galaxies could produce and release enormous quantities of helium and other materials into space by means of little bangs. Thus, conceivably, instead of one Big Bang, there were myriad little bangs throughout space and time, leading to a relatively stable cosmos punctuated by such explosions.

Hoyle decided to join with Fowler and Wagoner (who was then a postdoctoral researcher working with Fowler at Caltech) on a project exploring the production of light elements in the universe. In December 1965, Wagoner met Peebles at a conference in Miami, and they engaged in a friendly exchange about how such elements could be forged.[19] Peebles soon released several groundbreaking papers detailing how helium was produced in the hot primordial fireball. Wagoner, Fowler, and Hoyle pursued their own systematic study of how all the light elements (and isotopes) were formed, matching their results to those elements' observed abundance in space.

For that joint project—perhaps a tactical maneuver, or maybe because he was still trying to clarify his stance on the rapidly emerging developments—Hoyle remained conditionally supportive of the idea that the Big Bang produced the cosmic helium, while leaving the option open that it could have been little bangs instead. Hence the main thrust of the seminal paper "On the Synthesis of Elements at Very High Temperatures," by Wagoner, Fowler, and Hoyle, published in April 1967, involved element production in the Big Bang, with secondary discussion about the possible role of massive stars at the center of galaxies.

Wagoner later noted:

> Fred's main motivation for that paper was influenced by claims of the Russian astronomer Ambartsumian that explosions were occurring in the nuclei of galaxies. Fred hypothesized that these were due to massive stars, which he called "little bangs." That was why we extended the range of baryons per photon to include the much higher values appropriate to such objects. He needed those to make the lightest elements in a steady-state universe.
>
> To me, the most important result of our paper was to show that precisely those elements that could not be made in other ways (Li6, Be9, B10, B11 could be made by cosmic ray interactions with interstellar gas, and all the heavier elements by stars) could be made in the early universe (H, H2, He3, He4, Li7).[20]

While many astronomers found that three-author paper highly influential, they focused almost exclusively on its Big Bang predictions.[21] Peebles—expressing, perhaps, the general sentiment of the mainstream community—didn't think highly of the little bang idea. As he described it: "They had a scenario with little bangs in galaxies. I would characterize it as an argument of desperation."[22]

Although he never stated this openly, clearly Hoyle was rooting for the newly triumphant Big Bang theory to fail. Hoyle had invested so much time and effort into steady-state that it had become a major source of pride. He could never understand why it was discarded so quickly by the astronomical community. It lacked the philosophical baggage of a beginning of time, which he saw as a major flaw of the Big Bang. Occam's razor shaves away far-fetched solutions unless all simpler solutions are exhausted. Therefore, Hoyle argued, astronomers must rule out all possible alternative explanations for the massive production of helium and the appearance of background radiation before leaping to the extreme conclusion (in his mind) that the entire cosmos had a start date. Little bangs, to him, represented a reasonable compromise—especially in light of the discovery of ultra-energetic phenomena in space in the form of quasars.

Sri Lankan physicist and astrobiologist Chandra Wickramasinghe, who had completed his PhD on the subject of interstellar grains of graphite in 1960 under Hoyle's supervision, did not understand why cosmologists remained so adamantly opposed to the steady-state hypothesis, even in its modified form: "The insatiable appetite for denigrating Steady-State Cosmology, often with the flimsiest of evidence, that I witnessed in the 1960s still puzzles me. I cannot help thinking that the reasons have a deeply cultural basis. Without going into the technical details on either side of the argument, my own cultural predilection was for a steady-state universe of some kind. Such a cosmology is consistent with the philosophical world view that pervades the Indian subcontinent, and in particular it is in harmony with Buddhist traditions that are prevalent in Sri Lanka."[23]

GAMOW'S FINAL QUEST

Whereas Hoyle had great trouble sustaining interest in his cosmological ideas after a vast majority of the scientific community and the general public shifted to believing in the Big Bang, Gamow had a different but still imposing dilemma to face: reclaiming credit as the founder of the notion—in its first rigorous form—along with Alpher and Herman. In the major news stories regarding the CMBR, it was mainly Penzias and Dicke, as leaders of their teams, who were sought for interviews. The Dicke Gravity Group's articles cited some but not all of the relevant papers of Gamow, Alpher, and Herman. In those papers, Gamow believed, were ample predictions of cosmic background radiation, with a range of possible temperatures.

In a letter dated September 29, 1965, Gamow sent Penzias a brief list of temperature predictions he, Alpher, and Herman had made in various single-author and coauthored papers. He wanted to make sure that Penzias had an accurate record. Gamow concluded the letter with a poke at Dicke's reputation: "Thus, you see the world did not start with almighty Dicke."[24]

At least for a time, Gamow remained upset with Dicke. As Peebles recalled, "He did send us letters, some of them, in fact, bitter. 'Why don't you give me more credit for the hot Big Bang?'"[25]

Alpher recalled how upset Gamow was when he inadvertently voted yes for Dicke to be elected to the National Academy of Sciences (NAS), erroneously believing the similarly named Gerhard H. Diecke, a spectroscopist at Johns Hopkins, was the one being nominated.[26] Dicke was awarded NAS membership in May 1967.

It was Alpher and Herman, however, who did the bulk of the letter writing—which they would continue well beyond Gamow's death. As Alpher noted: "There is a whole seething caldron of unhappiness over the years, 1965 to date. . . . I have a file full of cases where we have written letters over the years, and it took us a long time to get to the point where we would even raise the issue with anybody. I still don't remember what triggered it. Why we ever wrote the first letter? Except maybe it was Gamow who pushed us."[27]

Some of Gamow's own letter writing took place in hospitals. Sadly, in his final years, he was no longer the strapping youth who had kayaked the Black Sea and motorcycled through Europe; rather, he was in very poor health. His excessive drinking, along with chain smoking and other unwise practices, had taken a great toll on his health.

His son Igor noted: "He was overweight. He had hardening of the arteries: arteriosclerosis. There was ringing in his ears. Got bad blood. Two years before he died, Father was actually institutionalized in Colorado Springs in a hospital. He stopped drinking. He walked around with a glass of ice water."[28]

Gamow's placement in a sanatorium in 1966 or early 1967 was something he agreed to because of a car crash when he had been drunk at the wheel. The harrowing experience woke him up to the fact that he needed to take major steps to turn his life around. He decided to become more serious about his health.

In January 1967, Gamow decided that it was safe to step back into the waters of astrophysics and cosmology by attending the Third Texas Symposium on Relativistic Astrophysics. Strangely enough, unlike the first two Texas Symposia (set in Dallas and Austin), the third was not situated in the Lone Star State at all but rather in the Big Apple—the Hotel New Yorker, to be specific, in association with the Goddard Institute for Space Studies.

It was a bustling place for Gamow to make his comeback. More than six hundred physicists from twenty-two different countries attended. Prominent attendees included Penzias, Wilson, Peebles, Wagoner, Sandage, Salpeter, Wheeler, Gold, and the Burbidges.

The mysteries of quasars and black holes were much on researchers' minds. Peebles and Wagoner each detailed their light element nucleosynthesis calculations. Repeated measurements of the CMBR were another focal point. Because of its size, with multiple sessions, participants noted that the atmosphere was very informal.

Gamow presided over a half-day session. But it was in between sessions, when a group was casually gathered around him, that he let out his feelings about the rediscovery of his Big Bang nucleosynthesis model and its prediction of cosmic radiation. He had "lost a penny" and Penzias and Wilson had "found a penny," he remarked. "Was it his penny?" Gamow asked.[29]

That year, Gamow also joined with Alpher and Herman on first joint cosmology paper in years, "Thermal Cosmic Radiation and the Formation of Protogalaxies," speculating how new galaxies formed while renewing their claim to the CMBR. Also, he contributed to a volume honoring Teller on his sixtieth birthday. Finally, encouraged

George Gamow, ailing in the hospital. CREDIT: AIP Emilio Segrè Visual
Archives, Physics Today Collection.

by his wife, Barbara, he decided to write his memoir, up to his fleeing
Russia for the United States in the 1930s. It would be published post-
humously, under Barbara's careful supervision and editing, with the
clever title *My World Line*. When asked about his later years (the 1940s
to the 1960s), Gamow would sometimes tell people, "Nothing happened to me interesting."[30]

Gamow's final research project was a reexamination of the large
numbers hypothesis of Dirac, with its prediction of changing fundamental constants, including the gravitational constant. That offered him an
opportunity to catch up with his old friend and engage in discussions by
mail. Unfortunately, during that period, autumn 1967 to summer 1968,
he was often incapacitated with medical issues, starting with a heart by-
pass procedure in fall 1967. By winter of that year, Gamow's liver started
to fail, which meant that he was often stuck in the hospital for days or
weeks at a time. His stamina to carry out the project was amazing, given
his repeated hospitalizations.

On December 2, 1967, Gamow wrote to Alpher from Saint Joseph's Hospital, where he had been staying for two weeks thus far, explaining both his medical diagnosis and his new passion for Dirac's theory.[31] He complained that his complexion had turned yellow because of hepatitis. Despite his discomfort, he emphasized, he felt driven to complete an article on the topic.

Dirac was a keen correspondent and sympathetic to Gamow's medical troubles. At one point, when Gamow was in the hospital, Dirac sent him a get well card seemingly intended for children, with an image of a tiger dressed like a doctor. It was unclear whether Dirac was being funny or actually thought it was appropriate to send a juvenile card to someone in his sixties with a critical condition.

Dicke and Gamow finally reconciled around that time on the basis of their mutual interest in the changing constants idea. Gamow seemed to forgive Dicke for overlooking his work. Dicke, in turn, was anxious to change the topic from the CMBR, which he had only marginal interest in, to his alternative cosmology ideas.

Gamow also began a correspondence with Freeman Dyson on the same topic. As Dyson recalled: "I knew George Gamow only at the end of his life, when we shared an interest in the question of possible variation of the constants of nature. He raised important questions to which nature gave negative answers. He was always respectful of nature's verdict. Unlike Hoyle who let his loyalty to a beautiful idea override the evidence."[32]

Gamow's last letter to Dirac is dated August 17, 1968. Gamow informed Dirac that he had reached the conclusion that the gravitational constant didn't change with time, after all. He wrote:

Dear Dirac,

I'm afraid that you will not like it. But may be [sic] you will accept it if it will be definitely shown that $\underline{G}$ does not change with time. But where to from here?

Yours as ever
George Gamow[33]

The graves of George and Barbara Gamow, Boulder, Colorado. CREDIT:
Photograph by Paul Halpern.

Dirac soon wrote back in defense of the changing constants idea
and the large numbers hypothesis, but it was too late. His friend, who
had brought him so much enjoyment, was gone forever.

On August 19, 1968, George Gamow's "world line"—the winding
thread linking Odessa, Saint Petersburg, Copenhagen, Cambridge,
Washington, Boulder, and other places—finally came to an end. He was
buried at Green Mountain Cemetery in Boulder, honored with a marble bench (that could be construed as resembling the Greek letter and
mathematical symbol pi) and a simple grave marker.

Gamow has continued to be honored by the two institutions where
he spent most of his academic career: George Washington University
and the University of Colorado (where his son Igor held a position as a
microbiology professor). In 1972, Gamow's colleagues at the University
of Colorado organized a memorial volume in his honor and invited contributors from around the world, including luminaries such as Dirac,
Teller, Stanislaw Ulam, Leon Rosenfeld, and Arno Penzias. Hoyle and

Plaque at George Washington University honoring George Gamow.
CREDIT: Photograph by Paul Halpern.

Narlikar contributed a piece, "Conformal Invariance in Physics and Cosmology," that, curiously enough, didn't mention Gamow at all, or even reference any of his works, as did many of the volume's other contributions.[34] Rather, it focused on a new set of gravitational equations that had mathematical properties that were different from Einstein's general theory of relativity. The omission was probably not an intentional slight, but it is likely that the two authors saw the volume as a chance to publish a work they had prepared, without considering its nature as a tribute.

In April 2000, George Washington University commissioned a plaque in Gamow's memory and had it placed in front of Samson Hall, where his office used to be. It honors him for his work on the Big Bang theory, nuclear physics, and popular communication, including the Mr. Tompkins series. Appropriately, it is near plaques honoring Bohr (for his 1939 nuclear announcement at the Fifth Washington Conference) and Teller.

After Gamow's death, Alpher (and to some extent Herman) continued to carry the flame for their work in nucleosynthesis. In a belated but well-deserved recognition, Alpher was awarded the National Medal of Science in 2005, "for his unprecedented work in the areas of nucleosynthesis, for his prediction that universe expansion leaves behind background radiation, and for providing the model for the Big Bang Theory."[35] He died in 2007.

Indeed, once a small field of research becomes an empire, sometimes its history is lost. By the time of Gamow's death, Big Bang cosmology had mushroomed into a thriving academic enterprise involving physics and astronomy departments around the world as well as numerous observatories. No wonder Alpher felt squeezed out.

Hoyle, in his later years, was excluded for a very different reason. The notion of accepting conventional views on the Big Bang, just because the majority supported it, was anathema to him. To strengthen his position, he founded the Institute of Theoretical Astronomy at Cambridge. But that, too, would eventually slip out of his hands, and then he truly was on his own—standing on the fringes of mainstream science.

CHAPTER NINE

Life on the Fringes

It was a very sad affair when Fred left Cambridge. He strongly supported British involvement in the AAT [Anglo-Australian Telescope] and quarreled with Ryle and the radio astronomers. In the struggle he lost—or gave up—directorship of the IOTA [Institute of Theoretical Astronomy] and went into self-imposed exile. We all ["Texas Mafia," Fred, and many others] rode a boat on the Cam and had one last party. I remember playing a guitar that Ardie Fowler rented for the occasion.

—Stan Woosley

I personally believe he would have done much better to leave Cambridge and go permanently to Caltech, but my mother refused to live in America. Pity because I believe the world lost my father's later years in science because of that fact. Living on the top of a mountain in the Lake District was perhaps not the best place to do active research.

—Elizabeth Jeanne Hoyle Butler

Astronomy at Cambridge has a venerable and complex history, as Hoyle was well aware. For much of his early career, he was one of the lone astrophysicists based in the applied math and theoretical physics contingent of the Faculty of Mathematics (scattered among various colleges), which in 1959 became organized as the Department of Applied Mathematics and Theoretical Physics (DAMTP). Other notables in that unit were Sciama and Dirac, and later Hawking.

Separate from that department was the Cambridge Observatory, where Eddington was director from 1914 until his death in 1944. Another notable director was John Couch Adams, one of the two nineteenth-century astronomers who correctly predicted the existence of Neptune (the other was French astronomer Urbain Le Verrier).

After the war, yet another astrophysics-related unit at Cambridge came into being: the radio astronomy group of Cavendish Laboratory. There, Martin Ryle reigned supreme, and those who disagreed with him were excluded.

Finally, connected with that group, the Mullard Radio Astronomy Observatory opened in 1957 in Lords Bridge, a few miles to the west of Cambridge. Ryle was heavily involved in planning that facility, partly funded by the electronics manufacturer Mullard Limited. There, Ryle constructed the 4C Array of radio telescopes with which he gathered the extensive data on radio sources that he asserted in 1961 was contradictory to steady-state.

By the mid-1960s, with much of the Cambridge astrophysics and cosmology community either hostile to Hoyle—such as Ryle—or supporters of the Big Bang—such as Sciama and Hawking—Hoyle started to feel squeezed out. He didn't have access to the computational resources he needed to continue his research. Moving permanently to the United States—in the footsteps of Gold and the Burbidges—was not an option because of Barbara's strong preference to stay in England. Therefore, he dedicated substantial energy to planning, and seeking funding for, a space where theoretical astrophysics and related topics could be discussed in an open-minded manner.

THE RISE AND FALL OF HOYLE'S INSTITUTE

Negotiations for a new center at Cambridge devoted to astronomy took several grueling years. In between meetings to discuss location, funding, staffing, and other details, Hoyle channeled his impatience into a very healthful aspiration: his goal was to climb all the nearly three hundred Munros in Scotland, mountains over three thousand feet in elevation. In that mission, conducted during numerous summer excursions to the Scottish Highlands throughout the late 1960s and early 1970s, Hoyle was often joined by Willy Fowler, his wife, Ardiane "Ardie" Fowler, and Fowler's prodigious student and professor at Rice University, Donald "Don" Clayton.

"I joined this theoretical movement in the late 1950s at Caltech as a research student of W. A. Fowler and evolved into a close friend of Hoyle," Clayton explained.[1]

During his hiking expeditions, Hoyle found the serenity to do some of his best thinking. He often carried a notebook—specifically, a type called a "jotter pad"—and pencil to record his thoughts. That's how many of his ideas for stories or calculations emerged.[2]

Another fun project Hoyle began around that period of his career was writing the libretti (sung dialogue) for two different pieces composed by American musician Leo Smit, a disciple of the acclaimed Russian composer Dmitri Kabalevsky. Smit also associated with Igor Stravinsky and interpreted the works of Aaron Copland, two other famous twentieth-century composers. Hoyle had met Smit in New York in 1953, and the two had hit it off and remained in touch. Smit once described Hoyle as having "a musical awareness that transcended many a professional musician's scope and understanding."[3]

By the mid-1960s, they had begun a kind of music–science dialogue. In 1965, Smit composed a short ditty in honor of Hoyle's fiftieth birthday. After Hoyle published his acclaimed science fiction novel *October the First Is Too Late* in 1966, Smit put together a recital of some of the music described in the book. He also put Hoyle's narrative to music in an oratorio about the Polish astronomer Nicolaus Copernicus. In 1969, their collaborative three-act comic opera, *The Alchemy of Love, or the Daemon Servant's Retribution*, premiered in New York.

By the time of that musical debut, Hoyle's bureaucratic struggles had been rewarded. In 1967, the grand opening of the Institute of Theoretical Astronomy (IOTA) fulfilled a long-standing dream. Hoyle quickly made it a base for transatlantic research with ties to the two American universities he had associations with: Caltech, because of Fowler; and Rice, because of Clayton. The Burbidges also visited Cambridge often, where they had long-standing ties.

Fowler and Clayton each spent considerable time at Cambridge, especially during summers (hence the hiking). Clayton, in turn, invited his students and fellow researchers from Rice and astrophysicists from other parts of Texas to join him in Cambridge each summer, a group that would be nicknamed the "Rice Mafia" or the "Texas Mafia."

Clayton described how that epithet arose: "Every summer research students from Rice U arrived in Cambridge. Hoyle provided office space and computer access to them. The other research staff and students started saying, 'The Rice Mafia has arrived.' The moniker stuck. The Rice Mafia did not carry its sinister overtones, but we did take over space and computer time for the aforementioned term. Our research publications (jointly from Rice U and IOTA) were numerous, causing Martin Rees to call these a 'golden five years of nucleosynthesis in Cambridge.'"[4]

One of the Rice Mafia members was Clayton's student, astrophysicist Stanford E. "Stan" Woosley, who remembered the warm friendship and camaraderie of the group, including many visits to pubs to chat and unwind after productive days doing research.[5] Yet another visitor was Wagoner, who spent two summers at the IOTA, motivated by Hoyle's energy and creativity. He was happy to take advantage of the IBM 360-44 mainframe computer situated in the institute's building, which offered him computational power to complete his calculations on nucleosynthesis. Hoyle invited him to his house several times to watch cricket on television and sip martinis.[6]

Hoyle's own astrophysics research turned to focus on the mystery of quasars as agents that injected new material and energy into the universe. He continued to hope quasars, as connected with active galactic nuclei, would offer a solution to the light element problem. If so, such little bangs could be a sensible alternative to the concept of a hot initial fireball, which he still found philosophically unsavory. Despite coauthoring several papers that described how the Big Bang could have created helium (and some lithium), the idea of an aging universe continued to disturb him.

There was a huge problem, though, with the concept of a perpetual, ageless universe based on element production in active galactic nuclei. All quasars had sizable redshifts, meaning that they were turbulent in the distant past (billions of years ago), not in the present. That disparity between the universe today and the universe back then strongly suggested that galaxies underwent an evolutionary process that made them calm down over time. For example, the massive core of the Milky Way does not spew enormous amounts of intense, high-energy radiation into space (otherwise we might be blasted). But perhaps billions of years ago it was active.

Controversially, Hoyle, working with Geoff Burbidge, tried to find alternative explanations for quasars' high redshifts that *didn't* involve Hubble expansion. Could, for example, the enormous gravitation of such massive energetic objects distort light through a process, first proposed by Einstein and later verified, called "gravitational redshifts"? If the bulk of quasars' redshifts was a gravitational effect, not an expansion effect, that would mean that they weren't moving away from us very fast. Hubble had shown that slow-moving galaxies (in terms of recessional speed) are much closer to us than fast-moving galaxies. That would place quasars much closer in distance and far more recent in time than thought. Hence, Hoyle and Burbidge argued, quasars are not an evolutionary effect but rather an ongoing part of how the universe continually renews itself.

By the start of the 1970s, more prosaic matters distracted Hoyle from his research. He became involved in planning the new Anglo-Australian Telescope in Australia, which involved a lot of traveling. But a much more onerous task was dealing with the bureaucracy at his home institution. The Cambridge administration didn't like the way astronomy at the university was divided into so many groups. While Cavendish remained hallowed ground because of the Rutherford legacy and the enormous productivity of Ryle's radio astronomy contingent, administrators put forth the proposal that the rest of Cambridge astronomy be reorganized into a single unit. That is, Hoyle's IOTA would merge with the observatory and form an Institute of Astronomy that covered the whole gamut of astronomy, including theoretical and observational. Although Hoyle liked the independence of IOTA, he decided to work with a university committee (including Ryle and some of his allies) to help plan the merger, including hiring new faculty and a director. The business got increasingly messy, with Hoyle feeling that his views were being shoved aside. Simon Mitton, a Cambridge astronomer and science historian who worked with both Ryle and Hoyle, witnessed this process and noticed that Hoyle increasingly believed that there was a conspiracy against him.[7]

When the majority of the committee, in opposition to Hoyle's opinion, selected theoretical astrophysicist Donald Lynden-Bell to be the first director of the united Institute of Astronomy, Hoyle was incensed. His dissent had little to do with Lynden-Bell himself, whom he liked and who would prove extremely capable. Lynden-Bell's groundbreaking research that showed how active galactic nuclei containing supermassive black holes provided the dynamos for quasars by gravitationally snatching nearby material meshed well with Hoyle's interests. But Hoyle had personal considerations. Age fifty-six at the time of the decision, he was close to the standard retirement age of sixty-seven. He dearly wished to spend his final few years at Cambridge working unhindered and shaping the future of its astronomy program, which meant having a loyal ally or himself as director. He decided that he would draw a line in the sand.

Either he would be named director of the Institute of Astronomy instead of Lynden-Bell, or he would resign from Cambridge.

On February 14, 1972, Hoyle composed a resignation letter, giving notice to the vice chancellor of Cambridge. Because of how Lynden-Bell's appointment was handled, he would step down from his position as Plumian Professor, a venerable professorship, on July 31, 1973.[8]

This put the university administration in a bind. Having Hoyle leave would be an embarrassment, but having him remain as a disgruntled director of the reformulated Institute of Astronomy for only a few years until his mandatory retirement would be an impediment. They kept the letter quiet for months as the process of Lynden-Bell's appointment continued to move forward. Meanwhile, Hoyle sent out feelers to see if other universities would offer him a position. Indeed, the University of Manchester, which maintained the famous Jodrell Bank Observatory, soon appointed him Honorary Research Professor of Physics and Astronomy. A second honorary appointment, at Cardiff University in Wales, arranged by Chandra Wickramasinghe, who worked there, would follow. Later in 1972, the prestigious honor of being knighted by Queen Elizabeth (making him "Sir Fred") offered him much-needed joy and recognition during that bleak period.

Although it had happened five decades earlier, there was still the young voice inside him that remembered the horror of being bullied and abused in Mornington School. Back then, he escaped by meandering along the lanes and byways of the Bingley region, finding solace in nature. This time, he and Barbara moved to an even remoter area, the Lake District of England, where they resettled at the start of 1973.

THE MOMENT OF REVELATION

In the days before the internet, location was an important part of doing science. An experimentalist needed to have access to a lab. Theorists, such as Gamow and Hoyle, required proximity to libraries, research centers, and perhaps computers. Ideally, they would be in a position to attend seminars and thereby keep up with the field. Phone and postal correspondence, the principal options for remote researchers (unless they had substantial travel budgets), were far less efficient means of staying up-to-date. By the time a remote theoretician received a journal in the mail and began to formulate ways to build upon a new result, those working at a major university with a well-stocked library had a jump start.

When Gamow, in what turned out to be his final decade of life, moved to Boulder, Colorado, in the foothills of the Rocky Mountains, he had sacrificed easy access to the universities and research centers of Washington, DC, and the relatively quick highway and train connections to numerous academic sites along the East Coast. True, the University of Colorado is a major institution, and he had full use of its resources. Yet he had given up proximity to the Applied Physics Laboratory of Johns Hopkins, the Carnegie Institution of Washington, and other centers with which he had collaborated—rendering him relatively isolated. The payoff for him was beautiful vistas and wide-open spaces. As a self-driven physicist who liked to follow his own path, he didn't seem to regret the move. Arguably, though, along with his health issues, it made him more marginal in the world of science during his final years.

Hoyle's relocation, though involving far fewer miles of travel, was an even more radical move. Rather than repositioning himself in another city or university town, he bought a two-hundred-year-old historic farmhouse on Cockley Moor, a wild area almost halfway up the northern slope of Helvellyn, one of the highest mountains in the Lake District and in all of England. Reaching the closest village, Dockray, involved walking or driving along a steep, winding road. Farther down, past scenic waterfalls, another winding road led to Ullswater, a pristine lake. It was (and still is) an extraordinary area for stargazing, but not so much for theoretical astrophysics collaborations.

Indeed, the mountain is best known as the abode of early nineteenth-century Romantic poets, such as Samuel Taylor Coleridge and William Wordsworth, rather than scientists. Their young contemporary, John

Keats, in his sonnet "Addressed to the Same," described it as a spiritual meeting place:

> Great spirits now on earth are sojourning;
> He of the cloud, the cataract, the lake,
> Who on Helvellyn's summit, wide awake,
> Catches his freshness from Archangel's wing

By moving to such a remote area, Hoyle cut the cord with the mainstream scientific community in a dramatic and resolute fashion. Isolated creativity—including rigorous mountain hikes with much time to think about the deep questions in science—was his top priority. No longer would he have to serve on committees and deal with red tape. Any collaborative efforts would be completely voluntary, involving only the closest friends he trusted. In short, he would be his own man.

The downside of the move soon became apparent. His serious astrophysics contributions, including collaborations with rising stars such as Clayton, gradually dropped off. Instead, he followed his hunches about broader questions in science, and in some cases these were areas where he had little expertise, for example, evolutionary biology. He'd bounce his ideas off of an ever-shrinking circle of friends and former students.

In 1973, a Yorkshire Television documentary, *Take the World from Another Point of View*, that focused on the research and philosophy of Richard Feynman spotlighted Hoyle's attitudes at the time. Feynman happened to be visiting the West Yorkshire village of Ripponden, in the picturesque South Pennines, where his third wife, Gweneth, was born and still had family. Yorkshire Television took the opportunity to interview him. As a famous Yorkshireman, Hoyle was invited to join Feynman for part of the filming. As they walked together through the village, they stopped by a historic pub, the Old Bridge Inn, where they chatted about science.

In one part, Feynman compared Hoyle's speculative attitude with his own cautious approach. He confessed that he was generally afraid to conjecture about nature because each time an idea came to mind, he often envisioned many alternatives. What if he was wrong? He noted that Hoyle, in contrast, seemed very comfortable guessing about the world.

Hoyle's response was telling. He remarked, "My choice is very simple. I don't set any requirement that the answer be right. It is just what I am interested in. That's the difference."

Explaining his own style, Feynman replied, "That's the difference. I am not trying to find out how nature *could be* but how nature *is*. See what's right. . . . Your idea is to find out what nature *could be*."

"No, no—what I think is *interesting*," said Hoyle.

"Even if it's wrong?" asked Feynman.[9]

Hoyle didn't reply. However, on the basis of that exchange one might view his continued support for variations of steady-state cosmology as a project that he found interesting rather than necessarily correct. Even if the Big Bang turned out to be a likely explanation, sticking up for a reasonable alternative seemed more compelling to Fred Hoyle. That is, he'd rather follow an engaging avenue of research that had at least some chance of being correct than a boring idea that had greater odds of being right.

Later in the dialogue, the two talked about "the moment of revelation": what happens during a scientific breakthrough. Hoyle likely was thinking about his remarkable discovery of the carbon-12 excited state when he described that experience: "You try all sorts of things, and you are hopeful about trying it—and you have a moment in a complicated problem when quite suddenly the thing comes into your head and you are almost sure that you have got to be right. . . . And then afterwards you wonder, now why the devil was I so stupid that I didn't see this."

Would more such moments of revelation be forthcoming? At the age of fifty-nine, Hoyle didn't know. But he believed that in his new environment at least he could pursue his own interests and speak his mind. That outspokenness would get him in trouble, though, on a number of occasions, including when he was publicly critical of a decision made by the Nobel Prize selection committee.

NOBEL OMISSIONS

In 1974, when the Nobel Prize in Physics was announced, Hoyle, along with much of the scientific community, was stunned by a glaring omission the Nobel committee had made. Martin Ryle and Antony "Tony" Hewish shared the honor for their work in radio astronomy, which, in Hewish's case, recognized the discovery of pulsars: radio sources that emitted signals with clockwork regularity. Yet it was well known that Hewish's graduate student, Jocelyn Bell, who had since changed her name to Bell Burnell upon getting married, first noticed the regular pulses in the collected signals back in 1967 when the discovery was made. Given her critical role in the pulsar findings, her being left out of the list of awardees seemed patently unfair.

The tale of Bell Burnell's encounter with the mysteriously rhythmic pulse from space had all the makings of a science fiction story. In fact, it strangely resembled the plot of Hoyle's own classic television series, A for Andromeda (co-developed by John Elliott), which had first aired on the BBC in 1961. In that speculative drama, which was turned into a novel, radio scientists discover a strange signal from the Andromeda galaxy that, because of its complexity, turns out to have been sent by a remote advanced civilization. Once they decipher its code, they discover that it offers the blueprints for a supercomputer, which, in turn, instructs them how to generate the embryonic form of an alien creature—which they name "Andromeda."

When Bell Burnell, who was familiar with Hoyle's tale and similar stories, found in her signal graphs drumbeat patterns from a distant radio source, she considered at first the possibility that it was communication from an alien race. When she brought up her concerns with her supervisor Hewish, they nicknamed the signal LGM1, standing for "Little Green Men." Hewish, worried about public reaction to the potential discovery of extraterrestrial messages, met with some of his colleagues to discuss strategy. Bell Burnell, meanwhile, looked through her data for similar signals, and she indeed identified some in another part of the sky. Discovering multiple sources settled the matter, given the unlikelihood of two civilizations, vastly remote from each other, sending similar messages. Hewish and Bell Burnell concluded that they were looking at a novel type of astronomical phenomenon, with publishable results and certainly worthy of further exploration. Each of their names appeared on all their papers on the topic. Before their first article appeared, Hewish gave a talk on the subject, and Hoyle was in the audience. Bell Burnell recalled Hoyle's insightful comments:

> When the paper had been accepted. . . . Tony gave a colloquium in the Cavendish in Cambridge, and we gave it a rather titillating title, you know, "A New Kind of Radio Source," or something like that. And word began to get around that something very interesting was happening, and people came in from everywhere. And I can remember that quite clearly. I can remember Fred Hoyle sitting in the front row. And at the end of colloquium Fred saying in his Yorkshire accent that, "This is the first I've heard about these things, but I don't think they're white dwarves. I think they're supernova remnants." In other words, he had taken in the fact that we found these very compact objects. For reasons that I don't recall why, he almost immediately could say he didn't think white dwarves were good candidates, but we were dealing with something to do with supernova—which of course is what it turned out to be.[10]

In the end, Hoyle proved to be absolutely correct. The strangely pulsing radio sources, called pulsars, turned out to be rapidly spinning neutron stars—the remnants of massive stellar cores after a supernova explosion.

"Fred, starting from cold, in forty-five minutes has hit the right explanation," noted Bell Burnell. "That man was a fantastic physicist. Really very impressive."[11]

Bell Burnell's savvy observations helped confirm a notion physicists had been discussing for years: the concept that for weighty stars undergoing catastrophic collapse of their core, gravity would trump the quantum exclusion rules that would otherwise stop their elementary particles from melding into a dense globe of closely packed neutrons.

When the Nobel Prize for Hewish was announced, some seven years after their discovery, Bell Burnell was extraordinarily gracious (and has continued to be for many decades). She was delighted that radio astrophysics itself was being recognized. Over much of the history of the Nobel Prize, trailblazing astronomers, such as Eddington, Slipher, Hubble, Baade, and so many others, have been ignored.

Bell Burnell happened to be one of the editors of *Observatory Magazine* at the time. The editorial group discussed congratulating Ryle and Hewish in one of the issues. Although she was absolutely supportive, some on the team told her that she had been treated unfairly, that the prize decision smacked of sexism and privilege. By excluding the young woman who actually did the work in favor of the older man who supervised her, it was truly a "No Bell" prize, they said.

About six months later, Hoyle, who was often interviewed because of his fame as a researcher, speaker, and writer, was asked his opinion of the prize being awarded to Ryle and Hewish. Hoyle let loose with a barrage of criticism aimed at the Nobel selection committee for overlooking Bell Burnell's key contribution to the discovery of pulsars. If he had stopped then and there, he would have voiced the view held by the majority of the scientific community. Unfortunately, though, emotionality and residual anger toward the Cambridge community—particularly the Cavendish radio astrophysics group—got the better of him. He conveyed to the journalists his belief that, because she wasn't present at certain meetings and talks, Hewish had tried to lock out Bell Burnell from sharing in her own discovery. Such press reports embarrassed and enraged Hewish, who, in fact, had listed Bell Burnell as first or second author on most of their joint papers. Hoyle, who was apparently worried about libel laws, ended up needing to correct the record by reemphasizing that his criticism had been aimed at the Nobel selection process rather than his former colleagues.

Flash forward to 1983, when the Nobel Prize was awarded for astrophysics once more, and once more stirred up another controversy. Again, the prize was split between two individuals. One-half went to Subrahmanyan Chandrasekhar for his work in understanding stellar structure and evolution. The other half went to Willy Fowler "for his theoretical and experimental studies of the nuclear reactions of importance in the formation of the chemical elements in the universe."

Of course, all of Hoyle's family members, friends, and colleagues knew that it was him who had pioneered the notion that all the chemical elements were created in stars. Hoyle had published his first paper on the topic in 1946, years before he had even met Fowler, and his second in 1954, also as a single-author work. As Clayton explained:

> Hoyle was the sole author of the papers establishing this theory, which steadily replaced the theory that those elements were created in an initial dense and hot condition in the universe (the Big Bang). Hoyle was very ahead of his time. His 1946 theory published in *Monthly Notices of the RAS* showed that the cores of evolving stars steadily approached the unsuspected high temperatures at which nuclei would assemble into the most stable nuclear species, which is the element iron. His second paper establishing his theory in 1954 in *Astrophysical Journal* showed that the stars would evolve through hotter and hotter shells, giving them an onionskin structure of increasingly heavier chemical elements, whose dispersal would naturally cause the increase in the abundances of the heavier chemical elements. "Nucleosynthesis happened in the stars" according to Hoyle.[12]

All of the members of the B²FH team—including Fowler himself—were stunned by the Nobel committee's omission. The team's highly influential paper of 1957 was a four-author project. Why in the world, they wondered, was Hoyle left out, and the Burbidges omitted as well?

Imagine if a pianist had assembled a musical quartet, coached his fellow musicians on the parts they played, and delivered an outstanding performance. Suppose the quartet recorded the concert, and critics loved it. Sometime later, let's imagine, on the basis of that recording the violinist received a Grammy Award for best new artist, as if the quartet's other members were backup players. They'd rightfully be fuming, especially the pianist who had orchestrated it all. That was Hoyle's situation.

Outwardly, Hoyle remained composed, however. He said very little about being overlooked in favor of his longtime friend, to whom he had introduced the topic, whereas his wife, Barbara, was far more vocal in expressing her outrage. Yet it clearly stung. As his son, Geoffrey, noted: "After thirty years of a very close friendship, he never had direct contact with Willy Fowler again."[13]

Pundits have speculated for years about why this happened. Only when the Nobel Prize nomination database for the year 1983, which includes materials related to the physics award decision, becomes public might the truth be known, but that won't be until at least 2033. The Nobel Foundation statute stipulates that material less than fifty years old must be kept private.[14] One reason why the whole B²FH team wasn't honored at once is clear: by tradition, a Nobel Prize can be awarded to only three individuals or organizations at a time. Honoring all four would have broken that rule.

"People didn't think it was fair," remarked Sarah Burbidge. "Every person in that foursome contributed equally. They couldn't give it to more than three people. But that's water under the bridge now."[15]

Perhaps Hoyle's critique of the 1974 decision played a role. Nobel selection committees have long memories. Many members are sensitive to prior accusations about their mechanisms, hidden as they are from the public view. Many worthy people, some have speculated, were accomplished enough to deserve the prize but didn't get it because of their vocal opinions. Along those lines, Narlikar conjectured that lingering hard feelings about Hoyle's outspokenness might have affected his chances.[16]

In discussing the matter with Fowler, Woosley learned his opinion that the committee had decided on Chandrasekhar first. Because Chandrasekhar was a theorist, they wanted the other half of the prize to go to an experimentalist, which ended up being Fowler. Hoyle would have been a second theorist.[17]

Another possible source of bias or confusion could have been the misconceptions of one of the nominators. Reportedly, Geoff Burbidge learned that Hans Bethe nominated Fowler because of the mistaken belief that he was the "team leader" of the B²FH project. Bethe found out only later that, though there was no real team leader, Hoyle was the driving force behind the original idea and had introduced the others to the topic. Once aware of the true sequence of events, Bethe allegedly expressed regret.[18]

As Geoff Burbidge noted in an article published in *Science*:

The theory of stellar nucleosynthesis is attributable to Fred Hoyle alone, as shown by his papers in 1946 and 1954 . . . and the collaborative work of B²FH. In writing up B²FH, all of us incorporated the earlier work of Hoyle. In my view, Hoyle's work has been undercited in part because it was published in an astrophysical journal, and a new one at that (the very first volume, in fact), whereas B²FH was published in a well-established physics journal, *Review of Modern Physics*. . . . Willy Fowler was very well known as a leader in that community, and the California Institute of Technology already had a news bureau that knew how to spread the word. . . . Hoyle should have been awarded a Nobel Prize for this and other work. On the basis of my private correspondence, I believe that a major reason for his exclusion was that W. A. Fowler was believed to be the leader of the group [which] was not the case.[19]

There is at least one other possible explanation. By the time the prize was awarded, Hoyle had started working on several projects with Wickramasinghe that had attracted great controversy. Among these were the concept of "panspermia," the idea that life came to Earth from deep space, and the related idea that diseases arrive periodically from space brought by small comets that break up in the upper atmosphere. Some pundits speculated that the Nobel committee didn't want to honor someone with fringe views.

The full story of why the Nobel committee overlooked Hoyle might never be known. If, in fact, it selected Chandrasekhar first, it would have had only two more slots open that year. The sensible move would have been to split the remainder of the award between Hoyle and Fowler rather than just the latter. But then there might have been complaints that two of the authors of the B²FH paper were overlooked, including a pioneering woman astronomer. If Bethe was indeed one of the nominators, and he was far more familiar with Fowler's work than that of the others, his views might have clinched the decision. Perhaps the eventual release of nomination letters and proceedings from 1983 will shed more light on the matter.

SUNSPOTS, DISEASES FROM SPACE, AND THE MISUNDERSTOOD ARCHAEOPTERYX

The Hoyle family was used to its brilliant patriarch advancing speculative opinions on a wide range of topics. He was a rebel and disinclined to follow any trends. And if he sometimes got lost on the wrong path through a dense forest of ideas, other times he would discover a remarkable shortcut to a brilliant clearing. His speculation, with no evidence at first, of that special carbon resonance level—that, when experimentally confirmed, had led to a revolutionary new theory of element production in stars—had been one of those cases. Even in Hoyle's final decades, there was always the hope that more such ideas would come. As his son, Geoffrey, remarked: "My father described himself as an observer of the world and ponderer on its problems. This curiosity combined with his mathematical knowledge and natural talent allowed his research to be far-reaching, covering all scientific disciplines, often to the scorn and derision of scientific contemporaries."[20]

Hoyle's new ventures didn't please all of his former colleagues, however. As Wagoner noted: "They were disappointed by his sometimes unfounded speculations."[21]

Like Gamow, Hoyle would venture into the realm of biology. However, there was a critical difference between their methodologies. Whereas Gamow always bounced his ideas off the "sounding boards" of expert biologists, such as Watson, and deferred to standard views of evolution and genetics, Hoyle veered much farther afield and consulted almost exclusively with Wickramasinghe, who was an expert in interstellar grains. Consequently, though Gamow's theories were not completely correct, mainstream geneticists continue to credit his insights. In Hoyle's case, his biological speculations have largely been met with sheer disbelief.

A persistent theme in Hoyle's work, dating back to his radio broadcasts of the 1950s, is that living organisms must be common in the universe. For example, a *Life* magazine profile of his work published in 1961 mentioned his views on the preponderance of living worlds: "Some scientists, such as Britain's Fred Hoyle, have been saying for years that evolution is not necessarily peculiar to earth. According to cosmologist Hoyle, there may be 100 billion planets with life on them in the Milky Way galaxy alone."[22]

From the late 1970s onward, usually in tandem with Wickramasinghe, Hoyle explored the notion, contrary to Darwin, that advanced life probably didn't have time to evolve fully on Earth, and thereby must have been seeded from elsewhere. For example, drawing on some of his speculative ideas in his novel *The Black Cloud*, interstellar clouds might have formed nurseries for living organisms to evolve slowly (perhaps even for billions of years, as time has no limits in a perpetually renewing universe) until a rogue comet transported such microbial forms of life into Earth's vicinity. If shards of such interstellar objects break up into Earth's atmosphere, alien microbes, called panspermia, could rain down and bring new, vibrant forms of life and, in some cases, devastating disease organisms.

By dismissing the Darwinian notion that life evolved from scratch on Earth, Hoyle attracted the interest of creationists. Given his earlier dismissal of organized religion, and his stated belief that the universe is eternal, it was an unusual match.

"Fred was far from being a religious man, not in the conventional sense at least," Wickramasinghe noted. "His position, as far as I could assess, was that *if* there were a cosmic creator it would be scarcely conceivable that any of the world's religions would have fully grasped either *His* intent or *His* plan."[23]

In 1981, on the basis of his book *Evolution from Space*, Hoyle was asked to be an expert witness for the state of Arkansas attesting that Darwinian evolution was not established fact. He was busy and sent Wickramasinghe to testify. Only later did Wickramasinghe learn about the "creation science" belief that the world is less than six thousand years old and regret participating in the trial.[24]

In tandem with his scientific pursuits, Hoyle continued to explore science fiction, often together with his son, Geoffrey. He was always delighted to become acquainted with fellow science fiction writers. On February 26, 1975, for example, during a research visit to Caltech, Hoyle had the pleasure of sharing the stage at the local YMCA with renowned writer Ray Bradbury for a literary discussion advertised as "The Promise of Science Fiction: Prophetic or Profane."

In December 1982, when Hoyle and Wickramasinghe were both attending a conference in Colombo, Sri Lanka, they met up with another famous science fiction author, Arthur C. Clarke, who was residing in that city at the time. Reportedly, the three chatted about the panspermia hypothesis, of which Clarke was supportive. "He was, of course, firmly on our side," recalled Wickramasinghe.[25]

For many of his former colleagues, Hoyle's panspermia speculations prompted raised eyebrows, but two outcrops of his main premise generated a far larger outcry. The first was the notion, expressed in his book *Diseases in Space* and related works, that all recent epidemics were caused by space microbes: from Legionnaire's disease to AIDS. Talking about microbial invasions in the distant past was one thing. Blaming tragic new diseases on extraterrestrial causes rubbed salt into raw wounds—as it seemed antithetical to the best scientific advice of the day. Related to that alien-induced illness concept was Hoyle and Wickramasinghe's bizarre letter to *Nature*, published in January 1990, claiming that sunspot cycles matched flu epidemics—which included the speculation that increased atmospheric electrical activity could conceivably compel viruses from space to plunge to Earth.

But even that was not Hoyle's most controversial claim. Arguably, the apex of his rogue statements on scientific issues was an assertion he made in 1985 that the Archaeopteryx, an internationally famous fossil of a bird-like dinosaur displayed at London's Natural History Museum, had been tampered with. The fossil offered a unique window into the transition between dinosaurs and feathered birds, but Hoyle alleged that someone, perhaps its discoverer, had made it look like the creature had feathers by systematically and fraudulently applying concrete to the cast. Antievolutionists rejoiced upon hearing Hoyle's allegation because it threatened to bring down the house of Darwin. The outcry forced the museum's scientists to complete a full investigation, leading to their published conclusion that no tampering had occurred and the fossil was absolutely real.

The following year, Hoyle and Wickramasinghe engaged in a more successful enterprise. They correctly predicted the appearance of a cometary nucleus based on their supposition that it contained organic materials.

THE RETURN OF THE COMET

In 1986, though Gamow was long gone, an emblem of his childhood returned. Halley's Comet blazed through the sky once again. In a curious juxtaposition, the expanding cosmos—the epitome of continuous growth—also loves cycles. The same object that tantalized little Geo would now prove an important test for Hoyle and Wickramasinghe's panspermia hypothesis.

Knowing that the European Space Agency (ESA) had launched the Giotto robotic probe to take pictures of the comet's nucleus, Hoyle and Wickramasinghe decided to offer a prediction about its appearance. They forecast that, because it contained organic material, the nucleus of the comet would be pitch black. Much to their delight, their prediction was proved correct.

On the night of March 13–14, Giotto passed within 370 miles of the comet and snapped a number of photographs of its nucleus. The flyby was historic. Never before had a comet been imaged so closely from space. Wickramasinghe described his and Hoyle's reactions to its findings: "The much publicised Giotto images of the nucleus of Comet Halley were obtained only after a great deal of image processing. The stark conclusion to be drawn from the Giotto imaging was the revelation of a cometary nucleus that was amazingly black. It was described at the time as being 'blacker than the blackest coal . . . the lowest albedo of any surface in the solar system. . . .' Naturally we jumped for joy! As far as we were aware at the time, we were the only scientists who made a prediction of this kind, a prediction that was a natural consequence of our organic/biologic model of comets."[26]

Although the Giotto probe confirmed Hoyle and Wickramasinghe's hypothesis about organic matter in comets, it didn't confirm their supposition about life coming from space via such objects. Most microbiologists and astrobiologists believe that bacteria and other microorganisms wouldn't be able to survive lengthy interstellar voyages, which would subject them to irradiation that would destroy them. Organic molecules can be made via chemical processes that don't involve life. Finding them in space doesn't provide conclusive evidence of living organisms.

The idea of panspermia, while still generally considered a fringe belief, pops up now and again. The 2017 discovery of Oumuamua, the first known interstellar object, brought the idea to the forefront once more. Could a rocky body convey microorganisms durable enough to survive millions of years in space? It is hard to see how, but perhaps there's a way. If so, Hoyle would be vindicated for one of his wildest ideas.

In 2018, Idan Ginsburg, Manasvi Lingam, and Abraham "Avi" Loeb, a team of researchers from Harvard, conducted a detailed study of the idea that interstellar bodies could transport microbes throughout the Milky Way. They concluded that "the entire Milky Way could potentially be exchanging biotic components across vast distances."[27]

THE ULTIMATE HORIZON

Along with his panspermia speculations, Hoyle continued to work in the 1980s and 1990s with Narlikar and Geoff Burbidge investigating alternatives to the Big Bang. (Margaret Burbidge had no interest in cosmology.) In numerous articles, he relayed his belief that the Big Bang had failed to live up to its promise, and that despite the cosmic microwave background revelations, it still was not a good match to observation. That was the theme of his 1982 Rede Lecture, "Facts and Dogmas in Cosmology and Elsewhere," reprinted in *The Sciences*, a journal published by the New York Academy of Science, under the title "The World According to Hoyle."

Alpher and Herman read the piece and were shocked by its mischaracterization of their predictions about the cosmic microwave background radiation. They felt that Hoyle had falsely claimed their early temperature estimates were wildly inaccurate by conflating them with some of Gamow's speculations, exaggerating the range of error in order to make his current explanation look more reasonable. They wrote an angry letter to the New York Academy conveying so much venom that the editors requested that it be toned down before publication. As Alpher recalled:

> The letter that we sent to the New York Academy in response to Hoyle's article had a final paragraph or so which rather hit hard at Hoyle, and the people at the New York Academy ([physicist and executive director] Heinz Pagels) prevailed on us to take it out . . . and we took it out. . . .
>
> What we said was that Hoyle was a guy who had done a hell of a lot of good work over the years, and we simply couldn't understand what he was doing now and for the last few years. . . . Hoyle is understandably bitter, I guess, because I take it he now has trouble getting his work published. He's now writing stuff for his books instead of journal articles.[28]

As the subject of a lot of criticism during that period, Hoyle remained sympathetic to the underdogs and mavericks in science. Science, he thought, should be open to a wider range of possibilities than the academy traditionally allowed. Needless to say, that opened him to correspondence from all manner of inexperienced researchers hoping for an eager ear. Remaining polite, he had a strategy for dealing with such circumstances.

As Stan Woosley recalled, "When aspiring cosmologists with little training and high regard for themselves wrote to him with unorthodox—even to him—theories of gravity and cosmology, he would introduce them to each other as distinguished colleagues and let them beat on each other rather than him."[29]

In 1988, Fred and Barbara Hoyle moved to the seaside resort of Bournemouth on the southern coast of England. Though, geographically, he wasn't as isolated as he was living in the Lake District, Hoyle's heterodox views continued to separate him from the mainstream scientific community. Nonetheless, wearing his nonconformity like a sacred garment, he pressed ahead with his cosmological musings.

A significant development in Big Bang cosmology helped convince Hoyle even more that he was on the right track: the rise of the inflationary universe model. The cosmological idea of *inflation*—a term coined by Alan Guth—is that the very early universe underwent an extraordinarily brief, ultrarapid burst of expansion, similar to the exponentially growing model of the universe proposed by de Sitter, before settling down into its current rate of growth, as described by Hubble's law. The inflationary expansion was so dramatic that, within a tiny fraction of a second, the region of space that is now the observable universe swelled from the merest subatomic speck, less than the size of a proton, into a sphere approximately the dimensions of a baseball. That spherical region then gradually grew at the much slower pace of a standard Friedmann-Lemaître-Robertson-Walker (FLRW) universe for more than thirteen billion years until it reached its current size.

The motivation for Guth and others (such as Russian physicist Andrei Linde) in proposing inflation had to do with several inexplicable features of the standard Big Bang theory. First, in what is called the "horizon problem," the standard Big Bang cannot explain how the average temperature of the CMBR, the distribution of galaxies, and other large-scale features are roughly the same in all directions, making the universe isotropic. We now know that in the history of the universe, at the time of recombination—when the first atoms were produced and the energy that became the CMBR was released into space—the overall universe must have already been coordinated in terms of average temperature and other properties. Yet the standard Big Bang dynamics do not predict that photons had time before then to cross the universe to even things out. Inflation offers that opportunity because it blows up a small patch, with equalized temperatures, so quickly that the evenness remains. It is like pouring hot coffee very rapidly into a large mug: the average temperature of coffee throughout the mug would immediately be the same, without the need to equalize over time—as opposed to very slow pouring in dribs and drabs, which might create temperature unevenness in the contents of the mug until it is stirred.

A second issue, called the "flatness problem," is that the present universe seems either perfectly flat, or very close to flat, in its spatial geometry. Yet, if one considers any irregularities in the early universe, such as overall curvature or local wrinkles, general relativity predicts that they would grow over time, not smooth out. Thus, in the standard Big Bang, only an extremely flat initial state of the universe would lead to its current flatness, or near flatness. Inflation opens up greater possibilities for the initial state. No matter how many wrinkles or how much curvature space starts off with, the era of rapid expansion would smooth it out, making it as flat as an ironed sheet.

The prime mover of the inflationary burst is a hypothetical source of energy called an inflaton. Some factor in the primordial universe switches on that mechanism, but only for a fleeting interval, allowing space to undergo exponential expansion akin to de Sitter's model. But, as Hoyle noted, it also resembled the steady-state model, similarly based on an energy field (the creation field) leading to continuous growth. Moreover, the inflationary model, like steady-state, doesn't suppose that all matter and energy instantly emerged at the dawn of creation. Rather, it proposes that a quantum decay process at the very end of the inflationary interlude, called "reheating," transformed energy into matter and released all the elementary particles into space. Although that is a sudden process, it is not quite the creation ex nihilo of the original Big Bang.

The fall of the original Big Bang, as Hoyle saw it, along with the earlier demise of the original steady-state model placed both early theories on similar ground and called for a completely new look at the suppositions of cosmology. Both visions are flawed, he argued. The original Big Bang, as advocated by Lemaître, Gamow, Alpher, Herman, and others, was based solely on the FLRW metric in which the universe underwent steady Hubble expansion driven solely by radiation and matter. But, as Hoyle pointed out, the inflationary hypothesis subverted that original picture and made the early universe—driven by an energy field—seem more like steady-state. Why would the inflationary era suddenly draw to a close, he wondered. If it didn't end, the universe would still contain a creation field, much like his own notion. Thus, he concluded, steady-state, with a perpetual creation field, was more natural than a brief burst of inflation that abruptly started, suddenly ended, and transformed into cosmic dynamics miraculously resembling the Big Bang.

Hoyle freely conceded that the original steady-state model—with element production based solely on stellar nucleosynthesis—failed to account for the helium abundance. Therefore, it required enormous energy bursts, on the scale of galaxies or larger, to produce the helium. That said, he argued, there was still a lot right about the model. It didn't have to face the horizon and flatness problems because of the way it expanded in a manner similar to de Sitter's cosmology rather than the FLRW-type models used in the Big Bang.

It was time, then, for a sensible weaving together of the various strands of cosmology. Hoyle suggested that inflation could take place continuously, on the scale of superclusters (large groupings) of galaxies, negating the need for a beginning of time. Of course, such a compromise would be a vindication of the gist of his original idea, which he continued to feel was unjustly dismissed.

That's where the quasi-steady-state universe, developed over many years but officially proposed by Hoyle, Narlikar, and Geoff Burbidge in 1993, came into play. It allowed for local centers of creation—the little bangs Hoyle had discussed in his earlier work—while the universe expanded indefinitely. Metallic cosmic needles, produced at the time of galaxy formation and made of either graphite or iron, would explain the microwave background. According to Hoyle's scheme, they served as "thermalizing agents" to absorb stellar radiation and reemit it at just the right microwave frequency distribution to replicate the CMBR temperature profile observed by astronomers.

Fellow astronomers beyond Hoyle's circle were generally so dubious of the metallic needles scheme that practically no one even took the time to address it. Contemporary internet searches turn up less than a hundred references to the hypothesis over the past few decades, which is an extraordinarily low number, given the enormous volume of research in modern astrophysics.

Peebles has pointed out a major problem with the idea. If needles scattered across the entire celestial dome produce thermal energy, they would block the signals of more distant radio sources.[30] Because we do in fact detect radio signals from extremely remote objects, the sky couldn't be pervaded with intervening needles that would mute them.

Increasingly precise mappings of the CMBR, including measurements taken by the Cosmic Background Explorer (COBE) satellite launched in 1989, pointed to almost identical temperatures in all directions of the sky, except for minute deviations from the norm here and there. The thermal profile, sometimes called the "baby picture of the universe," beautifully matched predictions for a hot Big Bang, followed by an era in which quantum fluctuations, stretched by cosmic dynamics, became the seeds of structure formation. That is, a completely bland picture would not have been able to explain how structures such as galaxies emerge. A completely turbulent picture, on the other hand, would not have supported a universal origin in the past. Rather, COBE measurements hit the sweet spot in between, revealing just enough deviation from the average to justify how space seems uniform on its largest scale but has many features, such as galaxies and clusters, on a slightly lesser scale.

Hoyle, Narlikar, and Burbidge persisted, nonetheless, in their critique of any variation of Big Bang cosmology. They argued that even the high-precision COBE measurements could be explained by smaller structures—little bangs—or local effects such as dust rather than by the entire universe in formation. Their shared ideas about the universe were summarized in the 1998 book *A Different Approach to Cosmology*. It was a labor of love and an expression of dissidence. By that point, with

Big Bang views so baked in, they didn't seem to expect that their thesis would sway mainstream thinking.

Shortly after Hoyle finished his contributions to that book, but before it was published, tragedy struck. On November 24, 1997, he was taking a walk around his native Gilstead after visiting his sister, Joan, who still lived in the family house. His scenic stroll took him to a rugged part of the border region between Gilstead and the nearby village of Eldwick, called Shipley Glen. The path followed the crest of a deep ravine. Suddenly, he lost his footing, perhaps on slippery leaves, and fell into a gulley. (There are speculations, owing to a missing wallet and other possible evidence, that he might have been pushed by an unseen assailant.) After assessing the situation, he tried to make his way back up to the main path, but slipped again, this time tumbling some three hundred feet down into the main ravine, where he landed on a bed of stones. He tried to get up but was severely injured and very disoriented. He would lie there, in freezing conditions, until a trained dog found him during a search and rescue mission.[31] Taken to a hospital in Bradford, he would physically recover, but his mind never quite felt the same. His only solace was that his contribution to his cosmology book had been completed before the accident.

From that point on, Hoyle's health declined. He and Barbara, who had been diagnosed by then with Parkinson's, were stymied by their own physical challenges. Unlike his perpetually renewing universe, his personal loss had no respite.

As the millennium of Copernicus, Galileo, Newton, and Einstein drew to a close, new discoveries in astronomy continued to proliferate. Foremost among those was the 1998 announcement by two teams of astronomers—the Supernova Cosmology Project and the High-Z Supernova Search—of evidence that the expansion of the universe is accelerating. That evidence derived from using a variety of exploding stars, called Type Ia supernovae, as standard candles, similar to how cepheids were employed by Henrietta Leavitt and others to gauge distances. By comparing their expected power output to their observed brightness, the groups could measure how far away such supernovae were. Then, by recording the Doppler shifts of the galaxies housing them, they could obtain their speeds and finally determine their acceleration. The result was most surprising. Virtually no one in the field of cosmology had anticipated an accelerating universe. Rather, they expected, based on the standard FLRW models, that the cosmos would be slowing down.

Naturally, the extraordinary discovery of the acceleration of cosmic expansion stimulated much discussion among theorists, in which Hoyle, because of his health challenges, could play only a minor role. He contributed to a five-author paper, "Possible Interpretations of the Magnitude-Redshift Relation for Supernovae of Type Ia," that attempted to match the supernova data with predictions of the quasi-steady-state model.

As a possible explanation of cosmic dynamics in light of the acceleration discovery, in early 2001, physicists Paul Steinhardt, Neil Turok, Burt Ovrut, and Justin Khoury proposed the "Ekpyrotic universe" (to be developed further by Steinhardt and Turok as the "cyclic universe"). In presenting the model, they also wanted to find an alternative to inflation. Making use of superstring theory and M theory—the notion that elementary particles are vibrating strands of energy—the researchers proposed that the universe underwent a collision in its past with another hyperplane separated from ours by an extra dimension. Their model resembles putting two slices of bread together to form a sandwich: whatever filling is placed between them naturally gets smoothed out. Hence, there would have been no need for an inflationary era. Moreover, as in the case of steady-state, the cosmos would have no beginning and no ending. Rather, it would cycle through various stages of creation and destruction, again and again. Although Hoyle loved to read about, and potentially criticize, alternatives to mainstream cosmology, it is doubtful that he learned about this model in the final months of his life.

On August 20, 2001, about a month and a half after his eighty-sixth birthday, Hoyle died of a stroke. So soon after the new millennium began, the world lost one of its most innovative minds, and cosmology would be a little less exciting for it. Alas, despite hopes for a steady-state, everything must end at some point.

Conclusion

THE LEGACIES OF GAMOW AND HOYLE

I believe that Father was one of the last great scientists to do science by the seat of his pants. Big science today, and particularly big physics, is done with large groups of people using magnificent and expensive machines. Nobody flies by the seats of their pants anymore.

But Father was first and foremost a storyteller. His scientific stories have been shared for more than a generation, including the one about the "Big Bang of the Universe."

—R. Igor Gamow, "Memories of My Father"

What I will remember about my father—for many things, but the thing I missed most when he died—was the fact that his wonderful mind would no more be having another of his mad ideas to discuss with me when I arrived to check up on my parents. What the world will remember, I suspect, is that he was the man who got Big Bang wrong because that is what the current crop of the great and the good in today's science will allow.

—Elizabeth Jeanne Hoyle Butler

In April 2007, almost six years after Hoyle's death, the American Physical Society held a session at its annual meeting honoring the fiftieth anniversary of the B^{2}FH paper on stellar nucleosynthesis.

Under the leadership of astrophysicist Virginia Trimble, the Program Committee of the Forum on the History of Physics, of which I was an active member, invited the Burbidges to speak. (Fowler had died in 1995.) Margaret and Geoff decided that Geoff would be the one to accept the invitation and fly to Jacksonville, Florida, where the conference was held. It was an honor for me to help arrange the talk.

I remember Geoff, who was then in a wheelchair and accompanied by a nurse, as a big man with a booming voice that reverberated throughout the packed lecture hall. His excellent talk was entitled "B^2FH, Nucleosynthesis and the Microwave Background." Afterward, I spent some time chatting with him. I asked him about his continued support of the quasi-steady-state model. His take is that cosmology needed alternatives. He compared unthinking acceptance of the Big Bang to mindless lemmings following their leader over a cliff.

Indeed, Geoff Burbidge offered a valid point. On occasion, mavericks and rebels drive new science. Sometimes their seemingly crazy hunches turn out to be absolutely correct. For example, Paul Dirac brilliantly predicted antimatter because of mathematical solutions he found to an equation he had developed. He could have just as easily dismissed those solutions as nonsensical. Instead, his prognostication proved correct and changed the course of physics.

In that context, it is clear that the intuitive, seat-of-the-pants styles shared by Gamow and Hoyle were absolutely needed in their time. With the science of element creation in a quagmire—not knowing how certain nuclear processes transpired—proceeding cautiously and incrementally might have led nowhere. It took the magnificent hunches of Gamow and Hoyle to leap over formidable gaps and move forward. Impulsive leaps involve a risk, no doubt, of getting it wrong, as Hoyle's late-life forays into fringe topics show. Nonetheless, at critical moments, the world needs divergent thinkers.

In the mid-twentieth century, cosmology and astrophysics required bold transformation. Imagine if every scientist of that day believed Einstein when he vehemently argued for a static universe and refused to interpret Hubble's data as evidence for expansion. Suppose they agreed with Einstein when he dismissed chance aspects of quantum physics. Without probabilistic quantum tunneling, there would be no satisfactory explanation for nuclear reactions in stars. We'd be stuck in the dead-end astrophysics and cosmology of the late nineteenth century, unable to explain how the sun and other stars produce their light, lacking an understanding of how the elements are built up, and oblivious to the dynamics and scale of the cosmos.

Gamow's and Hoyle's scientific contributions were revolutionary, reshaping our fundamental picture of the universe. Though they never directly collaborated, their research on the origin of the chemical elements complemented each other's in ways neither anticipated. In essence, one wrote the beginning of the story of element creation, and the other wrote the ending. Hoyle's theory could not explain why about one-quarter of the atomic content of the universe is helium, and Gamow's could not justify how elements such as carbon, nitrogen, oxygen, and dozens of even heavier ones exist. Together, though, their processes could explain everything under the sun—and beyond.

In recent years, laboratory simulations of conditions in the early universe have splendidly corroborated Gamow's notion of how the hot Big Bang forged enormous quantities of helium in only a few minutes. For example, in November 2020 a team of researchers working at Gran Sasso National Laboratory in Italy announced a precise measurement of the rate in which deuterium, when bombarded with protons, transforms into helium-3.[1] Remarkably, the group's result closely matched predictions of the primordial abundances of those isotopes found by analyzing the cosmic microwave background radiation. The prescience of Gamow's cosmological conjecture continues to astound.

Well before he ventured into cosmology, Gamow had already made headlines with his extraordinary contributions to nuclear physics. He brilliantly applied basic quantum mechanics to develop a simple but powerful explanation of how protons, alpha particles (helium nuclei), and other ions might tunnel through the formidable energy barrier associated with each nucleus, in both directions, thus explaining nuclear processes such as scattering, fusion, and emission. His methods enabled the systematic development of energetic accelerators that speed up particle projectiles and hurl them into targets, yielding a wealth of data about nuclear structure and processes. Gamow also suggested to Bohr the liquid drop model of the nucleus and worked with Houtermans and Atkinson to turn Eddington's hypothesis that stars shine because of hydrogen fusion into a more viable model of stellar nucleosynthesis of the light elements based on quantum processes.

Finally, later in life, Gamow's active mind turned to the subject of genetics. Although ultimately he didn't get it right, he made important suggestions to Watson and others about applying the methods of combinatorics. In that vein, he suggested the "triplet code" for amino acid production. His ideas help lead to discoveries in the genetic code that connect the patterns of nucleotides in DNA and RNA to the formation of various proteins.

Unfortunately, Hoyle's own late-life ventures into biology were not as memorable or lauded. His panspermia hypothesis would not have raised eyebrows if weren't for the fact that he argued that Earth's age of around 4.5 billion years was too brief for life to have evolved independently. In saying that, he seemed to be bashing Darwin and unintentionally aligning himself with the last people he would wish to be associated—fundamentalist creationists. For many scientists, Hoyle's stand on that issue placed him so far out in left field that they were hesitant, perhaps, to recognize his other accomplishments. (According to some pundits, it was one of the factors that lost him the support he needed to get a Nobel Prize.)

Often for political reasons, society doesn't treat kindly scientists of achievement who maintain views that are controversial. For example, in retrospectives of his work the fact that physicist Freeman Dyson broke with the overwhelming majority in the scientific community and expressed his opinion that human-caused climate change is not a problem often overshadows his life of accomplishment. Hoyle faced such criticism but with even greater venom.

Another burden weighing down Hoyle's legacy is the fact that the steady-state theory of the universe, if remembered at all, is treated as a misguided notion that never had a chance for success. Its proponents are sometimes treated like flat-earthers—those oblivious to reality. On the contrary, until astronomers got the timing of the Big Bang right by revising estimates of Hubble's constant and tacked on the hypothesis of a very early interlude of inflation (ultrarapid expansion), the Big Bang theory was in fact quite speculative, and exploring the steady-state alternative was perfectly reasonable until observational evidence in the early 1960s made it unworkable.

Mainstream astronomers did not look kindly upon Hoyle's continued efforts to tweak the theory into an increasingly contrived "quasi-steady-state" model. Still, many of those who knew Hoyle well continue to believe that science has gone too far in asserting the Big Bang as absolute truth and rejecting competitors. Hoyle's daughter, Elizabeth, remarked: "My personal sense is that a lot of things are coming to light in astronomy, but it is pretty damn boring to someone like myself, or indeed my grandchildren, because there is now no debate: the Big Bang is correct and anyone doubting the gospel is instantly sent into the outer darkness. I joke, of course, but my conversation with the so-called great and good around Cambridge has none of the excitement my father could generate, even in me!"[2]

Setting the various incarnations of steady-state and panspermia aside, Hoyle's career included so much to be honored and cherished. Astrophysicist Stan Woosley ranked Hoyle's four greatest scientific accomplishments, each of which was pivotal:

First, the idea that the heavy elements are made in supernovae. Of course, as is well known, he did this in the context of a steady-state model for the cosmos, needed to make the heavy elements in a universe where only hydrogen was created spontaneously. He promoted the idea, which was not entirely novel (Eddington and others in the 1920s talked about making heavy elements in stars), and is associated with it. His 1946 paper was seminal.

Second, the synthesis of iron in nuclear statistical equilibrium in supernovae. He had the details wrong. Iron is not made as ^{56}Fe (iron-56) but as ^{56}Ni (nickel-56), but the general idea was right. With Fowler, he proposed some of the early ideas about how a supernova explodes including rotation and nuclear burning. Both survive in modern models. He missed the importance of neutrino energy transport. [Astrophysicist Stirling] Colgate got that.

Third, predicting on the basis of stellar models the existence of an excited state at a specific energy in ^{12}C (carbon-12) that served as a resonance for helium burning. It was the prediction of this state and the validation at Kellogg Labs that got Fowler and Hoyle together.

Fourth, he and Fowler delineated the origin and mechanism for Type I and Type II supernovae: Type I from white dwarf explosions and Type II from massive star core collapse.[3]

I would add a fifth item to the list: Hoyle's contributions (along with Tayler, and later with Wagoner and Fowler) to the notion that the abundance of helium in the universe could not be explained by stellar processes and had to have emerged in a hot fireball. In making that case, he showed that he was independent-minded enough to be open to alternative models if they fit the data better. Once observational evidence pointed otherwise, he was willing to give up the idea that all the elements were made in stars.

In short, Hoyle transformed our scientific understanding of the late stages of stars, including the process of supernova explosions, as well as our comprehension of how the chemical elements came to be. Following the path Baade had pioneered, he rightly showed how the two stellar populations—Population I and II—are related, akin to how a new growth forest might take root in the rich soil of a former generation of trees devastated by fire. That's why Population I stars—and their planetary systems—are much richer in metals (heavy elements beyond helium).

In gauging the legacies of George Gamow and Fred Hoyle, we must look at not only their phenomenal scientific achievements but also their tremendous impact on scientific communication to the public. For them, the two aspects were deeply intertwined. By the time Gamow released *The Creation of the Universe* and Hoyle published *The Nature of the Universe* based on his BBC radio talks, it was clear that the audience for much of what they were doing in science was the public and posterity, as well as their fellow scientists. Each won the prestigious Kalinga Prize for the Popularization of Science—in 1956 and 1967, respectively—for his extraordinary ability to convey science to lay people. The emergence of modern science media, accessible to the broader population, not just specialists, made that possible.

Before the invention of the printing press, debates about far-reaching questions, such as whether time has a beginning, were confined to theologians and philosophers in exalted settings such as ecumenical councils. Few in the general public were privy to such discourse. Once books and journals became widely distributed, and scientific methods developed, scholars had far greater opportunity to argue their case in a manner that educated readers might fathom. Still, one would need to have specialized knowledge to appreciate the nuances of such discussions.

George Gamow and Fred Hoyle came of age during yet another transformation in how fundamental ideas might be discussed: the era of mass communications, including wide-reaching popular science media. Each loved Hollywood and the world of the fantastic. With Gamow sometimes picturing himself as a lone cowboy exploring the scientific Wild West, and Hoyle, a film noir sleuth unraveling clues in murder mysteries, each saw himself as a rebellious character standing larger than life. Gamow's motorcycling and Hoyle's long-distance hiking and mountaineering added to their images as solitary nonconformists akin to maverick heroes of the big screen. Consequently, each was able to convey scientific ideas with cinematic flair and a sense of wonder. They were vying for the attention of ordinary people, not just for the endorsements of scientists.

In June 1960, the *New York Times* ran a piece about the growing popularity of paperback books on science. The article noted how such inexpensive books made it possible for almost anyone to follow scientific debates: "One can even follow running controversies in paperbacks. An example of this is the controversy over whether the universe had a beginning or has simply 'always' existed, matter being continuously created. George Gamow, a proponent of the view that the universe was created in what he has called the 'big bang' or the 'big squeeze,' states his case in 'The Creation of the Universe.' . . . The opposing theory, championed by Fred Hoyle, can be found in his 'Frontiers of Astronomy.'"[4]

With accessible science literature, readers of the 1950s and 1960s could peruse a library or bookshop and find riveting, real-life accounts of extraordinary explorations of the depths of space and time, as fantastic as the popular science fiction of Isaac Asimov, Ray Bradbury, Robert Heinlein, and others. Kids could debate each other in the schoolyard about whether or not a blast from the past started everything, including homework, or if things were always pretty much the same, such as the cafeteria food. Many of those who read Gamow and Hoyle grew up to be scientists themselves.

In my case, *One, Two, Three . . . Infinity* was one of my favorite nonfiction books as a teenager, and *The Black Cloud* impressed me with its imaginative look at extraterrestrial life. Both scientists inspired me. Though I never had the opportunity to hear Gamow speak, I recall attending during my college days a very clear, well-organized, stimulating lecture by Hoyle, delivered unhurriedly in his West Yorkshire accent.

Hoyle and Gamow showed how accomplished scientists could write riveting speculative literature, bridging the gap between the "two cultures" decried by C. P. Snow. Hoyle's science fiction works, cowritten in some cases with his son, Geoffrey, stand the test of time for their imaginative look at space travel, time travel, aliens, and other mind-blowing concepts. Lines such as "Our consciousness corresponds to just where the light falls, as it dances about among the pigeon holes [of time],"[5] from Hoyle's classic novel about temporal turmoil, *October the First Is Too Late*, are as profound and imaginative as any full-time, professional science fiction writer has written. Since his father's death, Geoffrey has admirably continued the Hoyle tradition with his own speculative fiction.

Although Gamow's Mr. Tompkins series, for which Snow was the editorial midwife (by inviting him to contribute to *Discovery* magazine), doesn't fall rigidly into the category of science fiction, it is certainly speculative literature based on science. The situations the title character finds himself in, such as a world in which light speed is much lower and matches the pace of common vehicles, are truly inventive. Impressively, Igor Gamow has also maintained his family's tradition by extending the series with new episodes that address more recent scientific questions, such as exploring black holes with Stephen Hawking.

Arguably, Hoyle's and Gamow's special mixture of factual popularization and fictional speculation helped inspire many in the next generation of scientist-writers—such as Carl Sagan, Alan Lightman, Janna Levin, Brian Greene, and Kip Thorne (each of whom has penned fiction as well as science)—to navigate the same balance successfully.

Not that Hoyle and Gamow were the first to combine the two. In the seventeenth century, German mathematician Johannes Kepler, a prolific writer of scientific works, also wrote an early science fiction novel, *Somnium*. *Alice's Adventures in Wonderland* by Lewis Carroll, the nom de plume of mathematician Charles Dodgson, who wrote serious works under his real name, serves as another example. That said, it wouldn't be until the age of broadcast media and paperbacks—information conduits mastered by Hoyle and Gamow—that the concept really took off. Therefore, we owe them thanks for not just enlightening the public but also inspiring subsequent writers to do so as well.

Sadly, because of Hoyle's and Gamow's important legacies as science communicators, some of their colleagues seem to undervalue their groundbreaking scientific contributions. They contend, perhaps because of jealousy, that being an outstanding science communicator as well as an excellent scientist is somehow a contradiction. An old joke holds that "those who can't do, teach." Arguably, the two thinkers' images as colorful, quirky popularizers set them up for their major accomplishments to be overlooked when the major accolades in their fields were bestowed—Hoyle's well-deserved knighthood being the notable exception.

Hoyle worried that Sir Harold Spencer Jones, the Astronomer Royal, saw him as a popularizer of steady-state rather than as its co-discoverer. More devastatingly, the Nobel Prize selection committee somehow overlooked the fact that he was the first one to propose that the chemical elements emerged from lighter elements via ultra-high-temperature stellar evolutionary processes, such as core contraction and supernova bursts. He introduced Fowler to his ideas, yet it was Fowler, not Hoyle, who was awarded for the discovery. In Gamow's case, the co-discoverers and interpreters of the discovery of the cosmic microwave background radiation were very familiar with his popularizations (for instance, Bob Wilson loved the Mr. Tompkins series when he was a child), but not so much with his theory of Big Bang nucleosynthesis.

In recounting the accomplishments of Gamow and Hoyle, let's not forget their brilliant collaborators who made much of their work possible. Some of them, such as Landau, Teller, and Fowler, are famous in their own right. Many have gone on to extremely successful careers, such as Clayton, Wagoner, and Narlikar (who remains a leading cosmologist in India). In the case of Alpher and Herman, while they also thrived in their positions, they were understandably upset that their important prediction of the microwave background radiation either went unnoticed or was misinterpreted (credited to Gamow).

Geoff Burbidge retained his fighting spirit until the end, even in his waning years when he faced considerable health problems. He died on January 26, 2010. Margaret lived a decade longer and died, at the age of one hundred, on April 5, 2020.

In 2018, in my role as chair of the Historic Sites Committee of the American Physical Society, I helped organize a historic plaque ceremony at Princeton University honoring the research of Bob Dicke and his group. In planning that event, it was a pleasure to get to know Jim Peebles, who is friendly, knowledgeable, generous, and down-to-earth. At dinner, after the ceremony, I sat at a table with Jim, his wife, Alison, and Bob Wilson. I asked them the details of the story of the famous phone call in which Dicke said, "we've been scooped," and delighted in hearing Jim and Bob recall the details. I'm sure they've told the tale countless times, but it was extraordinary to hear it in person.

I called Peebles on August 26, 2019, to interview him for this book. What a wonderful surprise when, little more than six weeks later, Professor Göran K. Hansson, secretary general of the Swedish Royal Academy of Sciences, announced that he was the co-recipient of the 2019 Nobel Prize in Physics. Dicke (who died in 1997) would have been extremely proud, no doubt.

Wilson was thrilled to see his old friend join the ranks of Nobel laureates. "I was extremely pleased that Jim Peebles got the prize."[6]

In his Nobel Lecture, delivered in Stockholm on December 8, Peebles offered a fascinating personal chronicle of the history of modern physical cosmology. Interestingly, he took time to acknowledge the contributions of Gamow and Hoyle, offering his take on their strengths and weaknesses, along with mentioning the legacy of Dicke in experimental cosmology:

> At the end of the Second World War, great energy was released into arts and sciences. In particular, after the war, three remarkable individuals started thinking about cosmology. . . . Of the three, George Gamow was, by far, the most brilliant, intuitive physicist I've ever met, but along with that intuition was a distinct lack of interest in following up ideas. In 1948, he laid out many basic ideas of our

standard, hot Big Bang cosmology, but he didn't pursue them. His ideas died away—had to be rediscovered. Fred Hoyle had brilliant ideas about a steady-state universe, but he had so liked that idea, he had trouble letting it go. . . . Bob Dicke's group remains strong and active to this day.[7]

Peebles's recognition, "for theoretical discoveries in physical cosmology," did not honor just a single finding but rather a career of accomplishments, including major contributions to what is often called the concordance model of cosmology, also known as λCDM (lambda–cold dark matter) cosmology, with λ symbolizing the cosmological constant. Because of the Nobel Prize–winning discovery in 1998 by two teams of astronomers that the Hubble expansion is speeding up, cosmologists posited a new kind of substance, called dark energy, that acts as a kind of antigravity to accelerate universal expansion. The properties of dark energy are such that researchers have found it useful, so far, to model it by reintroducing the cosmological constant term Einstein discarded in the early 1930s. Hence the λ. Cold dark matter is a different kind of hypothetical substance that emits no discernible light or other form of electromagnetic radiation, yet makes itself known through its gravitational attraction to visible bodies. As Peebles and others demonstrated, galaxies and other large-scale features of the universe would be unstable without cold dark matter helping bind them together through their hidden gravitational strength. The "cold" part stems from the realization that hot particles would interact more fleetingly and therefore would not be able to serve as a kind of astral glue. (Think of what would happen if you heated up household glue and sprayed it haphazardly over two objects you wished to cement together versus applying room-temperature glue slowly and carefully to the objects' surfaces to be joined, for greater effect.)

The dynamics of the model are thereby governed by the flat geometry solution of Einstein's equations, filled with a mixture of cold dark matter, ordinary matter, and radiation, with the addition of a cosmological constant. In essence, it is an open Big Bang cosmology that will expand forever (and never collapse), with its expansion speeding up.

Following in the footsteps of Slipher and Hubble, astronomers have been testing the predictions of the λCDM cosmology throughout the early twenty-first century. Some of those observations pertain to the visible objects in space, such as supernovae in distant galaxies, while others are high-precision profiles and detailed statistical analyses of data collected by satellite probes of the cosmic microwave background radiation. It still isn't clear, however, whether all the cosmological parameters being measured (current value of the Hubble constant, cosmic acceleration parameter, percentages of dark matter and dark energy, and so forth) can be tweaked to exactly match a concordance model. Meanwhile, theorists with alternative cosmological ideas are waiting in the wings, hoping evidence will reveal gaps in λCDM that might be explained by their own theories.

As Peebles recently remarked: "Surely there are more adjustments to come. An example may be the current 10 percent discrepancy in the rate of the universe's expansion derived in two different ways . . . supernova measurements [and the] cosmic microwave background. Maybe the difference is down to a subtle systematic error, which wouldn't be surprising for these difficult measurements. Or maybe it is evidence for something new. I haven't joined the search for what that something new may be, but I will be fascinated to see what people come up with."[8]

Furthermore, despite numerous theories, nobody knows yet what the dark energy and cold dark matter actually are. Tests using sensitive detectors continue in labs around the world, with hopes of finding glimpses of such substances.

The friendly, competitive spirit of cosmology persists. In today's world, with the study of the universe continuing to reveal new surprises, George Gamow and Fred Hoyle would feel right at home. Long live their passions for scientific explanation and their vibrant quests for cosmic discovery!

Acknowledgments

Thanks to the faculty, administration, and staff of University of the Sciences for their continued encouragement, including Paul Katz, Jill Baren, Vojislava Pophristic, Elia Eschenazi, Jessie Taylor, and the other members of the Department of Math, Physics, and Statistics.

Many thanks to Igor and Elfriede Gamow, Elizabeth Jeanne Hoyle Butler, Geoffrey Hoyle, Sarah Burbidge, P. James E. Peebles, Robert W. Wilson, Arno A. Penzias, Kenneth C. Turner, Robert Wagoner, Jayant Narlikar, Donald Clayton, Stanford E. Woosley, Wendy Teller, Joanne Page, Cormac O'Raifeartaigh, Nicholas Booth, George Pothering, and the late Freeman Dyson for sharing their fascinating recollections. Thanks also to Nancy Dicke Rapoport, John Rapoport, Chandra Wickramasinghe, Adam Crothers, Kathryn McKee, Anita Hollier, Virginia Trimble, Alan Chodos, David Cassidy, Brian Keating, Lyman Page, Bruce Partridge, Rainer Weiss, Ashutosh Jogalekar, Joseph Martin, Diana Kormos-Buchwald, Alberto Martinez, and the late E. Margaret Burbidge.

I deeply appreciate the preceding research in the history of cosmology and all of the libraries and archives that offered me assistance with this project, including St. John's College Library, Cambridge;

Princeton University Library Special Collections Rare Books Division; the George Washington University Special Collections Research Center; the Library of Congress; the Niels Bohr Archives in Copenhagen; and the CERN Archive in Meyrin, Switzerland.

I offer my sincere appreciation to Elfriede and Igor Gamow, representing the Gamow Estate, for permission to quote from George Gamow's correspondence, and to Nancy Dicke Rapoport for permission to quote from Robert H. Dicke's correspondence.

Many thanks to the extraordinary editorial staff at Basic Books, including TJ Kelleher and Eric Henney, and to my wonderful agent, Giles Anderson of the Anderson Literary Agency.

Thanks to my friends for their encouragement, including Fred Schuepfer, Pam Quick, Michael Erlich, Mari Errico, Simone Zelitch, Doug Buchholz, Lisa Tenzin-Dolma, Lindsey Poole, Greg Smith, Frank Cross, Mitchell and Wendy Kaltz, Mark Singer, Michal Meyer, Bob Jantzen, Boris Briker, and Kris Olson. Above all, my deep appreciation to my family for their support and advice, including Stanley Halpern, Arlene Finston, Eli Halpern, Thessaly McFall, Aden Halpern, and Felicia Hurewitz.

Further Reading

Alpher, Victor S. "Ralph A. Alpher, George Antonovich Gamow, and the Prediction of the Cosmic Microwave Background Radiation." *Asian Journal of Physics* 23, nos. 1 and 2 (2014): 17–26.

Bartusiak, Marcia, ed. *Archives of the Universe: 100 Discoveries That Transformed Our Understanding of the Cosmos.* New York: Vintage, 2006.

———. *Black Hole: How an Idea Abandoned by Newtonians, Hated by Einstein, and Gambled On by Hawking Became Loved.* New Haven, CT: Yale University Press, 2015.

———. *The Day We Found the Universe.* New York: Pantheon, 2009.

———. *Thursday's Universe: A Report from the Frontier on the Origin, Nature, and Destiny of the Universe.* New York: Times Books, 1986.

Chown, Marcus. *The Afterglow of Creation.* Herdon, VA: University Science Books, 1996.

———. *The Magicians: Great Minds and the Central Miracle of Science.* London: Faber & Faber, 2020.

Cline, Barbara Lovett. *The Questioners: Physicists and the Quantum Theory.* New York: Crowell, 1965.

Ellis, George. *Before the Beginning: Cosmology Explained.* New York: Boyers/Bowerdean, 1993.

Farmelo, Graham. *The Strangest Man: The Hidden Life of Paul Dirac, Mystic of the Atom.* New York: Basic Books, 2009.

Ferris, Timothy. *The Whole Shebang: A State-of-the-Universe Report*. New York: Simon & Schuster, 1997.

Gamow, George. *The Creation of the Universe*. New York: Viking, 1952.

———. *My World Line: An Informal Autobiography*. New York: Viking, 1970.

———. *Thirty Years That Shook Physics: The Story of Quantum Theory*. New York: Doubleday, 1966.

Gough, Douglas, ed. *The Scientific Legacy of Fred Hoyle*. New York: Cambridge University Press, 2005.

Greenstein, George. *Portraits of Discovery: Profiles in Scientific Genius*. New York: John Wiley & Sons, 1998.

Gregory, Jane. *Fred Hoyle's Universe*. New York: Oxford University Press, 2005.

Halpern, Paul. *Einstein's Dice and Schrödinger's Cat: How Two Great Minds Battled Quantum Randomness to Create a Unified Theory of Physics*. New York: Basic Books, 2015.

———. *Time Journeys: A Search for Cosmic Destiny and Meaning*. New York: McGraw-Hill, 1990.

Harrison, Edward. *Cosmology*. New York: Cambridge University Press, 1981.

Hawking, Stephen. *A Brief History of Time: From the Big Bang to Black Holes*. New York: Bantam Books, 1988.

Hoyle, Fred. *Home Is Where the Wind Blows: Chapters from a Cosmologist's Life*. Herndon, VA: University Science Books, 1994.

Hoyle, Fred, Geoffrey Burbidge, and Jayant V. Narlikar. *A Different Approach to Cosmology—from a Static Universe Through the Big-Bang Towards Reality*. New York: Cambridge University Press, 2000.

Keating, Brian. *Losing the Nobel Prize: A Story of Cosmology, Ambition, and the Perils of Science's Highest Honor*. New York: W. W. Norton, 2018.

Kragh, Helge. *Cosmology and Controversy: The Historical Development of Two Theories of the Universe*. Princeton, NJ: Princeton University Press, 1999.

———. *Masters of the Universe: Conversations with Cosmologists of the Past*. New York: Oxford University Press, 2015.

Lightman, Alan, and Roberta Brawer. *Origins: The Lives and Worlds of Modern Cosmologists*. Cambridge, MA: Harvard University Press, 1990.

Livio, Mario. *Brilliant Blunders: From Darwin to Einstein—Colossal Mistakes by Great Scientists That Changed Our Understanding of Life and the Universe*. New York: Simon & Schuster, 2013.

Mack, Katie. *The End of Everything (Astrophysically Speaking)*. New York: Scribner, 2020.

McConnell, Craig S. "The Big Bang–Steady-State Controversy: Cosmology in Public and Scientific Forums" (PhD diss., University of Wisconsin, Madison, 2000).

Misner, Charles W., Kip S. Thorne, and John A. Wheeler. *Gravitation*. San Francisco: W. H. Freeman, 1973.

Mitton, Simon. *Fred Hoyle: A Life in Science*. New York: Cambridge University Press, 2011.

Peebles, P. James E. *Cosmology's Century: An Inside History of Our Modern Understanding of the Universe*. Princeton, NJ: Princeton University Press, 2020.

Peebles, P. James E., Lyman A. Page, and R. Bruce Partridge, eds. *Finding the Big Bang*. Cambridge: Cambridge University Press, 2009.

Penrose, Roger. *The Road to Reality*. London: Jonathan Cape, 2004.

Reines, Frederick, ed. *Cosmology, Fusion & Other Matters: George Gamow Memorial Volume*. Boulder: Colorado Associated University Press, 1972.

Segrè, Gino. *Ordinary Geniuses: Max Delbrück, George Gamow, and the Origins of Genomics and Big Bang Cosmology*. New York: Viking, 2011.

Seife, Charles. *Alpha & Omega: The Search for the Beginning and End of the Universe*. New York: Viking, 2003.

Silk, Joseph. *The Big Bang: The Creation and Evolution of the Universe*. New York: W. H. Freeman and Company, 1980.

Singh, Simon. *Big Bang: The Origins of the Universe*. London: Fourth Estate, 2005.

Smoot, George, and Keay Davidson. *Wrinkles in Time*. New York: William Morrow, 1993.

Teller, Edward, with Judith Shoollery. *Memoirs: A Twentieth-Century Journey in Science and Politics*. Cambridge, MA: Perseus, 2001.

Tyson, Neil deGrasse, and Donald Goldsmith. *Origins: Fourteen Billion Years of Cosmic Evolution*. New York: W. W. Norton, 2004.

Watson, James. *Genes, Girls, and Gamow: After the Double Helix*. New York: Oxford University Press, 2001.

Weinberg, Steven. *The First Three Minutes: A Modern View of the Origin of the Universe*. New York: Basic Books, 1977.

Wheeler, John Archibald, with Kenneth W. Ford. *Geons, Black Holes, and Quantum Foam: A Life in Physics*. New York: W. W. Norton, 2000.

References

INTRODUCTION: THE QUEST FOR THE ORIGIN OF EVERYTHING

1. Elizabeth Jeanne Hoyle Butler, personal communication with the author, September 5, 2019.

2. Fred Hoyle, "Continuous Creation," *The Listener* 41 (April 7, 1949): 568.

3. Geoffrey Hoyle, personal communication with the author, September 30, 2019.

4. "Ralph Alpher and Robert Herman—Session II," interview by Martin Harwit, August 12, 1983, American Institute of Physics, Niels Bohr Library and Archives, Oral Histories, https://www.aip.org/history-programs/niels-bohr-library/oral-histories/3014-2.

5. Robert V. Wagoner, personal communication with the author, November 1, 2019.

6. Virginia Trimble, "Obituary, E. Margaret Burbidge (1919–2020)," *Nature*, April 27, 2020, https://www.nature.com/articles/d41586-020-01224-9.

7. Jeremy Bernstein, *Nuclear Weapons: What You Need to Know* (New York: Cambridge University Press, 2008), 193.

8. C. P. Snow, "The Two Cultures" (Rede Lecture, Cambridge University, Cambridge, England, May 7, 1959).

CHAPTER ONE: CHILDREN OF AN EXPANDING COSMOS

1. George Gamow, *My World Line: An Informal Autobiography* (New York: Viking, 1970), 9–10.

2. "Comet's Poisonous Tail," *New York Times*, February 8, 1910.

3. "Comet Notes," *Scientific American*, May 21, 1910, 416.

4. Stephen Castle, "Yes This Is Britain's Happiest Place," *The Independent*, November 29, 2017, https://www.independent.co.uk/news/long_reads/britains-happiest-place-craven-skipton-yorkshire-dales-a8065771.html.

5. Geoffrey Hoyle, personal communication with the author, September 30, 2019.

6. Fred Hoyle, *Home Is Where the Wind Blows: Chapters from a Cosmologist's Life* (Herndon, VA: University Science Books, 1994), 26–27.

7. Elizabeth Jeanne Hoyle Butler, personal communication with the author, September 5, 2019.

8. Simon Mitton, *Fred Hoyle: A Life in Science* (New York: Cambridge University Press, 2011), 20.

9. Fred Hoyle, interview by Alan Lightman, August 15, 1989, American Institute of Physics, Niels Bohr Library and Archives, Oral Histories, https://www.aip.org/history-programs/niels-bohr-library/oral-histories/34366.

10. Hoyle, *Home Is Where the Wind Blows*, 52–53.

11. Fred Hoyle, interview by Alan Lightman, August 15, 1989, https://www.aip.org/history-programs/niels-bohr-library/oral-histories/34366.

12. Joseph Conrad, *Under Western Eyes* (London: Methuen, 1911), 184.

13. "Revolution in Science: New Theory of the Universe: Newtonian Ideas Overthrown," *Times of London*, November 7, 1919, 1.

14. George Gamow, interview by Charles Weiner, April 25, 1968, American Institute of Physics, Niels Bohr Library and Archives, Oral Histories, https://www.aip.org/history-programs/niels-bohr-library/oral-histories/4325.

15. Hoyle, *Home Is Where the Wind Blows*, 69.

16. Geoffrey Hoyle, personal communication with the author, September 30, 2019.

17. Geoffrey Hoyle, personal communication with the author, September 30, 2019.

CHAPTER TWO: PREPARING THE BATTLEFIELD

1. George Gamow, interview by Charles Weiner, April 25, 1968, American Institute of Physics, Niels Bohr Library and Archives, Oral Histories, https://www.aip.org/history-programs/niels-bohr-library/oral-histories/4325.

2. Alexander Friedmann to Albert Einstein, December 6, 1922, Einstein Archives, https://einsteinpapers.press.princeton.edu/vol13-trans/363.

3. Albert Einstein, "Note to the paper by A. Friedmann, 'On the Curvature of Space,'" *Zeitschrift für Physik* 16 (June–July 1923): 228.

4. George Gamow, *My World Line: An Informal Autobiography* (New York: Viking, 1970), 45.

5. Allan Sandage, interview by Paul Wright, May 16, 1974, American Institute of Physics, Niels Bohr Library and Archives, Oral Histories, https://www.aip.org/history-programs/niels-bohr-library/oral-histories/32874.

6. Mario Livio, "Mystery of the Missing Text Solved," *Nature* 479 (2011): 171–173.

7. Gamow, *My World Line*, 44.

8. Cormac O'Raifeartaigh and Brendan McCann, "Einstein's Cosmic Model of 1931 Revisited: An Analysis and Translation of a Forgotten Model of the Universe," *European Physical Journal H* 39 (2014): 63–85.

9. Cormac O'Raifeartaigh and Simon Mitton, "Einstein's Oxford Blackboard: A Unique Historical Artefact" (unpublished manuscript).

10. Cormac O'Raifeartaigh, "Einstein's Steady-State Cosmology," *Physics World*, September 2014, 1.

11. Cormac O'Raifeartaigh, Brendan McCann, Werner Nahm, and Simon Mitton, "Einstein's Steady-State Theory: An Abandoned Model of the Cosmos," *European Physical Journal H* 39 (2014): 1.

12. Fred Hoyle, *Home Is Where the Wind Blows: Chapters from a Cosmologist's Life* (Herndon, VA: University Science Books, 1994), 152.

CHAPTER THREE: UNLOCKING THE NUCLEUS

1. George Gamow, *Thirty Years That Shook Physics* (Mineola, NY: Dover, 1985), 51.

2. R. Igor Gamow, phone interview by the author, September 2, 2019.

3. Barbara Lovett Cline, *The Questioners: Physicists and the Quantum Theory* (New York: Crowell, 1965), 127–128.

4. R. Igor Gamow, phone interview by the author, September 2, 2019.

5. Freeman Dyson, personal communication with the author, February 22, 2019.

6. George Gamow, quoted in Cline, *The Questioners*, 129.

7. R. Igor Gamow, phone interview by the author, September 2, 2019.

8. R. Igor Gamow, phone interview by the author, September 2, 2019.

9. Graham Farmelo, *The Strangest Man: The Hidden Life of Paul Dirac, Mystic of the Atom* (New York: Basic Books, 2009), 172.

10. J. J. Thomson, "Ions Mine," http://ww3.haverford.edu/physics-astro/songs/ionsmine.htm.

11. R. Igor Gamow, personal communication with the author, April 10, 2020.

12. Dennis Christiansen, "The Bohr Legacy," University of Copenhagen, https://uniavisen.dk/en/the-bohr-legacy/.

13. George Gamow, interview by Charles Weiner, April 25, 1968, American Institute of Physics, Niels Bohr Library and Archives, Oral Histories, https://www.aip.org/history-programs/niels-bohr-library/oral-histories/4325.

14. R. Igor Gamow, phone interview by the author, September 2, 2019.

15. George Gamow, interview by Charles Weiner, April 25, 1968, https://www.aip.org/history-programs/niels-bohr-library/oral-histories/4325.

16. R. Igor Gamow, phone interview by the author, September 2, 2019.

17. Farmelo, *The Strangest Man*, 259.

18. R. Igor Gamow, phone interview by the author, September 2, 2019.

19. R. Igor Gamow, personal communication with the author, June 12, 2020.

20. Rho Gamow, quoted in Edward Teller with Judith Shoollery, *Memoirs: A Twentieth-Century Journey in Science and Politics* (Cambridge, MA: Perseus, 2001), 124.

21. R. Igor Gamow, phone interview by the author, September 2, 2019.

22. Teller and Shoollery, *Memoirs*, 124.

23. George Gamow, quoted in Teller and Shoollery, *Memoirs*, 139.

24. Fred Hoyle, *Home Is Where the Wind Blows: Chapters from a Cosmologist's Life* (Herndon, VA: University Science Books, 1994), 131.

25. Geoffrey Hoyle, personal communication with the author, September 30, 2019.

26. "Fred Hoyle: A Singular Life," https://www.hoyle.org.uk/A-Singular-Life/.

27. Geoffrey Hoyle, personal communication with the author, September 30, 2019.

CHAPTER FOUR: RECURRENCE IN THE DEAD OF NIGHT

1. Fred Hoyle, "An Assessment of the Evidence Against the Steady-State Theory," box 3, folder 2, St. John's College Library, Papers of Sir Fred Hoyle, Cambridge.

2. Geoffrey Hoyle, personal communication with the author, September 30, 2019.

3. Elizabeth Jeanne Hoyle Butler, personal communication with the author, September 5, 2019.

4. Elizabeth Jeanne Hoyle Butler, personal communication with the author, September 5, 2019.

5. H. H. Hopkins to Fred Hoyle, May 10, 1948, box 87, folder 9, St. John's College Library, Papers of Sir Fred Hoyle, Cambridge.

6. Harold Spencer Jones to Fred Hoyle, June 27, 1952, box 87, folder 9, St. John's College Library, Papers of Sir Fred Hoyle, Cambridge.

7. Cormac O'Raifeartaigh, personal communication with the author, June 12, 2020; Cormac O'Raifeartaigh and Simon Mitton, "Interrogating the Legend of Einstein's 'Biggest Blunder'" (draft manuscript).

8. Helge Kragh, "What's in a Name: History and Meanings of the Term 'Big Bang,'" Cornell University, January 2013, https://arxiv.org/abs/1301.0219.

9. Helge Kragh, "Big Bang: The Etymology of a Name," *Astronomy & Geophysics* 54, no. 2 (April 2013): 2.28–2.30, https://doi.org/10.1093/astrogeo/att035.

10. Geoffrey Hoyle, personal communication with the author, September 30, 2019.

11. Ralph Alpher, interview by Martin Harwit, August 12, 1983, American Institute of Physics, Niels Bohr Library and Archives, Oral Histories, https://www.aip.org/history-programs/niels-bohr-library/oral-histories/3014-2.

12. George Gamow, *My World Line: An Informal Autobiography* (New York: Viking, 1970), 126–128.

13. Albert Einstein to Jean-Jacques Fehr, 1952, Albert Einstein Archive Online, Archive no. 26-074, quoted in Cormac O'Raifeartaigh, Brendan McCann, Werner Nahm, and Simon Mitton, "Einstein's Steady-State Theory: An Abandoned Model of the Cosmos," *European Physics Journal H* 39 (2014): 353–367.

14. Wolfgang Pauli to Aniela Jaffé, December 3, 1951, trans. and reprinted in Carl Jung and Wolfgang Pauli, *Atom and Archetype — The Pauli/Jung Letters, 1932–1958,* ed. C. A. Meier, trans. David Roscoe (Princeton, NJ: Princeton University Press, 2001), 71.

15. Fred Hoyle, *Home Is Where the Wind Blows: Chapters from a Cosmologist's Life* (Herndon, VA: University Science Books, 1994), 310.

16. Simon Mitton, *Fred Hoyle: A Life in Science* (New York: Cambridge University Press, 2011), 131.

17. Geoffrey Hoyle, personal communication with the author, September 30, 2019.

18. Jayant Narlikar, personal communication with the author, August 2, 2019.

19. Fred Hoyle, *The Nature of the Universe* (Oxford: Blackwell, 1950), 125.

20. Geoffrey Hoyle, personal communication with the author, September 30, 2019.

21. Geoffrey Hoyle, personal communication with the author, September 30, 2019.

CHAPTER FIVE: ALPHA TO OMEGA

1. Ralph Alpher, interview by Martin Harwit, August 11, 1983, American Institute of Physics, Niels Bohr Library and Archives, Oral Histories, https://www.aip.org/history-programs/niels-bohr-library/oral-histories/3014-1.

2. R. Igor Gamow, "Memories of My Father," in *Mr. Tompkins Gets Serious: The Essential George Gamow*, ed. Robert Oerter (New York: Pi Press, 2005).

3. Ralph Alpher, interview by Martin Harwit, August 11, 1983, https://www.aip.org/history-programs/niels-bohr-library/oral-histories/3014-1.

4. Ralph Alpher, interview by Martin Harwit, August 11, 1983, https://www.aip.org/history-programs/niels-bohr-library/oral-histories/3014-1.

5. Ralph Alpher, interview by Martin Harwit, August 11, 1983, https://www.aip.org/history-programs/niels-bohr-library/oral-histories/3014-1.

6. "Bottle, Cointreau, YLEM" object, Smithsonian Institution, National Air and Space Museum, Washington, DC, https://www.si.edu/object/nasm_A20010321000.

7. Cormac O'Raifeartaigh and Simon Mitton, "Interrogating the Legend of Einstein's 'Biggest Blunder,'" *Physics in Perspective* 20 (2018): 318–341.

8. George Gamow to Oskar Klein, April 1948, Niels Bohr Archives, Copenhagen, Denmark, by permission of the Gamow Estate.

9. Oskar Klein to George Gamow, April 1948, Niels Bohr Archives.

10. P. James E. Peebles, phone interview by the author, August 26, 2019.

11. Robert V. Wagoner, personal communication with the author, November 1, 2019.

12. George Gamow to Ralph Alpher, with a note to Robert Herman, 1956, George Gamow Papers, Library of Congress, by permission of the Gamow Estate.

13. George Gamow to Ralph Alpher, summer 1948, George Gamow Papers, Library of Congress, by permission of the Gamow Estate.

14. Ralph Alpher, "Johns Hopkins Science Review," March 18, 1949, quoted in Paulette Campbell, "Applied Physics Lab Pioneer Receives National Medal of Science" (press release), Johns Hopkins University, July 23, 2007, https://www.jhuapl.edu/PressRelease/070723.

15. Anthony Turkevich to Ralph Alpher and Robert Herman, October 24, 1949, Arno Penzias Papers, Library of Congress.

16. Stephen Maran and A. G. W. Cameron, "Relativistic Astrophysics," *Science*, September 27, 1967, 1517.

17. Pope Pius XII, quoted in José G. Funes, *"The Vatican Meeting on Stellar Populations,"* in *The Vatican Observatory, Castel Gandolfo: 80th Anniversary Celebration*, ed. Gabriele Gionti and Jean-Baptiste Kikwaya Eluo (New York: Springer, 2018), 199.

18. "Pope Says Science Proves God Exists," *New York Times*, September 8, 1952.

19. J. Richard Gott III (@JRichardGott), "In his basement, Gamow had a whole wall of books," Twitter, August 23, 2011, 9:04 p.m., https://twitter.com /JRichardGott/status/106200251328770051.

20. R. Igor Gamow, phone interview by the author, September 2, 2019.

21. R. Igor Gamow, phone interview by the author, September 2, 2019.

22. R. Igor Gamow, phone interview by the author, September 2, 2019.

23. Ralph Alpher, interview by Martin Harwit, August 11, 1983, https:// www.aip.org/history-programs/niels-bohr-library/oral-histories/3014-1.

24. R. Igor Gamow, phone interview by the author, September 2, 2019.

25. George Gamow, interview by Charles Weiner, April 25, 1968, https:// www.aip.org/history-programs/niels-bohr-library/oral-histories/4325.

26. Fred Hoyle, "The Big Bang in Astronomy," *New Scientist*, November 19, 1981, 522.

CHAPTER SIX: BUILDING THE ELEMENTS

1. A member of the "RNA Tie Club," along with Watson, Gamow, Feynman, and others, Delbrück would serve in the 1960s as Igor Gamow's postdoctoral supervisor.

2. Elizabeth Jeanne Hoyle Butler, personal communication with the author, September 5, 2019.

3. "Obituary of Raymond Lyttleton," *Times of London*, May 19, 1995, 19.

4. Donald Clayton, personal communication with the author, August 11, 2019.

5. C. W. Cook, W. A. Fowler, C. C. Lauritsen, and T. Lauritsen, "B12, C12, and the Red Giants," *Physical Review* 107 (1957): 508.

6. Geoffrey Hoyle, personal communication with the author, September 30, 2019.

7. E. Margaret Burbidge, interview by David Devorkin, July 13, 1978, American Institute of Physics, Niels Bohr Library and Archives, Oral Histories, https://www.aip.org/history-programs/niels-bohr-library/oral-histories/25487.

8. John D. Hillaby, "Cosmic Creation Held Continuing: British Astro-Physicists Say Elements in Universe Are Being Steadily Formed," *New York Times*, September 8, 1954.

9. Harrison Smith, "Margaret Burbidge, Who Explored the Universe's Distant Reaches, Dies at 100," *Washington Post*, April 8, 2020.

10. Geoffrey Burbidge, "Hoyle's Role in B^{2}FH," *Science* 319 (March 14, 2008): 1484.

11. Sarah L. Burbidge, phone interview by the author, November 3, 2018.

12. Sarah L. Burbidge, phone interview by the author, November 3, 2018.

13. Geoffrey Hoyle, personal communication with the author, September 30, 2019.

14. "Physicist Makes Helium of Carbon; Transmutation Is Hailed as Helping to Explain Origin of Universe; 'Big Bang' Theory Hit," *New York Times*, December 31, 1956.

15. Geoffrey Hoyle, personal communication with the author, September 30, 2019.

16. Pope Pius XII to Father Daniel O'Connell, oral communication, Archive of the Vatican Observatory, quoted in José G. Funes, "The Vatican Meeting on Stellar Populations," in *The Vatican Observatory, Castel Gandolfo: 80th Anniversary Celebration*, ed. Gabriele Gionti and Jean-Baptiste Kikwaya Eluo (New York: Springer, 2018), 201.

17. Georges Lemaître, quoted in Daniel J. K. O'Connell, ed., *Stellar Populations: Proceedings of the Conference Sponsored by the Pontifical Academy of Science and the Vatican Observatory, May 20–28, 1957* (New York: Interscience Publishers, 1958), 284.

18. Elizabeth Jeanne Hoyle Butler, personal communication with the author, September 5, 2019.

19. Fred Hoyle, *Home Is Where the Wind Blows: Chapters from a Cosmologist's Life* (Herndon, VA: University Science Books, 1994), 301.

20. Geoffrey Hoyle, personal communication with the author, September 30, 2019.

21. Fred Hoyle, *The Black Cloud* (London: William Heinemann, 1957), 200.

22. Geoffrey Hoyle, personal communication with the author, September 30, 2019.

23. P. James E. Peebles, phone interview by the author, August 26, 2019.

24. Ralph Alpher, interview by Martin Harwit, August 11, 1983, American Institute of Physics, Niels Bohr Library and Archives, Oral Histories, https://www.aip.org/history-programs/niels-bohr-library/oral-histories/3014-1.

25. R. Igor Gamow, phone interview by the author, September 2, 2019.

26. Simon Mitton, *Fred Hoyle: A Life in Science* (New York: Cambridge University Press, 2011), 169.

27. Elizabeth Jeanne Hoyle Butler, personal communication with the author, September 8, 2019.

28. Alan Cottey, "Martin Ryle: An Energy Visionary," *Physics World*, September 11, 2018, https://physicsworld.com/a/martin-ryle-an-energy-visionary/.

29. Hoyle, *Home Is Where the Wind Blows*, 270.

30. Martin Ryle, "Halley Lecture," Oxford 1955, quoted in "History," Cavendish Astrophysics, University of Cambridge, https://www.astro.phy.cam.ac.uk/about/history.

31. Chandra Wickramasinghe, *A Journey with Fred Hoyle* (Singapore: World Scientific, 2013), 14.

32. Hoyle, *Home Is Where the Wind Blows*, 410.

33. Robert Jastrow, quoted in John W. Finney, "Satellite Challenges Theory of Universe," *New York Times*, December 20, 1961, 13.

CHAPTER SEVEN: TRIUMPH OF THE BIG BANG

1. Robert H. Dicke to Peter Franke, October 23, 1963, Robert H. Dicke Papers, Special Collections, Princeton University Library, by permission of Nancy Dicke Rapoport. "Scalar component" refers to a type of energy field that is variable from point to point rather than constant over space and time.

2. "Cosmology: Math Plus Mach Equals Far-Out Gravity," *Time*, June 26, 1964.

3. Robert H. Dicke to Arthur P. Moor, May 27, 1965, Robert H. Dicke Papers, Special Collections, Princeton University Library, by permission of Nancy Dicke Rapoport.

4. P. James E. Peebles, phone interview by the author, August 26, 2019.

5. Robert H. Dicke to Stephen W. Hawking, October 26, 1965, Robert H. Dicke Papers, Special Collections, Princeton University Library, by permission of Nancy Dicke Rapoport.

6. Robert H. Dicke, interview by Martin Harwit, June 18, 1985, American Institute of Physics, Niels Bohr Library and Archives, Oral Histories, https://www.aip.org/history-programs/niels-bohr-library/oral-histories/4572.

7. Arno Penzias, phone interview by the author, October 18, 2019.

8. Robert W. Wilson, phone interview by the author, October 17, 2019.

9. Robert W. Wilson, phone interview by the author, October 17, 2019.

10. Robert W. Wilson, phone interview by the author, October 17, 2019.

11. Robert W. Wilson, phone interview by the author, October 17, 2019.

12. Arno Penzias, phone interview by the author, October 18, 2019.

13. Arno Penzias, phone interview by the author, October 18, 2019.

14. Robert W. Wilson, phone interview by the author, October 17, 2019.

15. Kenneth C. Turner, personal communication with the author, October 18, 2019.

16. P. James E. Peebles, phone interview by the author, August 26, 2019.

17. Robert W. Wilson, phone interview by the author, October 17, 2019.

18. Robert W. Wilson, phone interview by the author, October 17, 2019.

19. Robert W. Wilson, phone interview by the author, October 17, 2019.

20. P. James E. Peebles, phone interview by the author, August 26, 2019.

21. P. James E. Peebles, quoted in Tom Garlinghouse, "A 'Joy Ride' of a Career: Peebles Wins Nobel Prize in Physics for Tackling Big Questions About the Universe" (press release), Princeton University, October 8, 2019, https://www.princeton.edu/news/2019/10/08/joy-ride-career-peebles-wins -nobel-prize-physics-tackling-big-questions-about.

22. P. James E. Peebles, phone interview by the author, August 26, 2019.

23. Kenneth C. Turner, personal communication with the author, October 18, 2019.

24. P. James E. Peebles, phone interview by the author, August 26, 2019.

25. P. James E. Peebles, phone interview by the author, August 26, 2019.

26. P. James E. Peebles, "How Physical Cosmology Grew," Nobel Lecture, December 8, 2019, https://www.nobelprize.org/prizes/physics/2019/peebles /lecture/.

27. Robert Dicke, interview by Martin Harwit, June 18, 1985, https://www .aip.org/history-programs/niels-bohr-library/oral-histories/4572.

28. P. James E. Peebles, Lyman A. Page, and R. Bruce Partridge, eds., *Finding the Big Bang* (Cambridge: Cambridge University Press, 2009), 192.

29. Robert W. Wilson, phone interview by the author, October 17, 2019.

30. Walter Sullivan, "Signals Imply a 'Big Bang' Universe," *New York Times*, May 21, 1965, 1.

31. Robert W. Wilson, phone interview by the author, October 17, 2019.

32. P. James E. Peebles, phone interview by the author, August 26, 2019.

CHAPTER EIGHT: THE POINT OF NO RETURN

1. Dennis Sciama, interview by Spencer Weart, April 14, 1978, American Institute of Physics, Niels Bohr Library and Archives, Oral Histories, https:// www.aip.org/history-programs/niels-bohr-library/oral-histories/4871.

2. Dennis W. Sciama to Robert H. Dicke, December 21, 1966, Robert H. Dicke Papers, Special Collections, Princeton University Library, by permission of Nancy Dicke Rapoport.

3. Roger Penrose, *The Road to Reality* (London: Jonathan Cape, 2004), xiv.

4. P. James E. Peebles, Lyman A. Page, and R. Bruce Partridge, eds., *Finding the Big Bang* (Cambridge: Cambridge University Press, 2009), 196.

5. C. P. Gilmore, "World's Greatest Mystery," *Popular Science*, November 1965, 104.

6. Robert H. Dicke to C. P. Gilmore, August 16, 1965, Robert H. Dicke Papers, Special Collections, Princeton University Library, by permission of Nancy Dicke Rapoport.

7. Peebles, Page, and Partridge, *Finding the Big Bang*, 196.

8. P. James E. Peebles, personal communication with the author, September 13, 2019.

9. G. J. Ringer to Robert H. Dicke, July 30, 1965, Robert H. Dicke Papers, Special Collections, Princeton University Library, by permission of Nancy Dicke Rapoport.

10. George Pothering to Robert H. Dicke, November 22, 1965, Robert H. Dicke Papers, Special Collections, Princeton University Library, by permission of George Pothering.

11. George Pothering, personal communication with the author, September 11, 2019.

12. George Pothering, personal communication with the author, September 11, 2019.

13. Freeman Dyson, personal communication with the author, February 22, 2019.

14. Robert W. Wilson, phone interview by the author, October 17, 2019.

15. Sarah L. Burbidge, phone interview by the author, November 3, 2018.

16. Engelbert Schucking, "The First Texas Symposium on Relativistic Astrophysics," *Physics Today*, August 1989, 52.

17. Robert V. Wagoner, personal communication with the author, November 1, 2019.

18. George Gamow, quoted in Bob King, "Twinkle, Twinkle, Quasi-Star," *Sky and Telescope*, April 3, 2019, https://skyandtelescope.org/observing/12 -quasars-for-spring-evenings/.

19. Peebles, Page, and Partridge, *Finding the Big Bang*, 196.

20. Robert V. Wagoner, personal communication with the author, November 1, 2019.

21. Stan Woosley, personal communication with the author, August 25, 2019.

22. P. James E. Peebles, phone interview by the author, August 26, 2019.

23. Chandra Wickramasinghe, *A Journey with Fred Hoyle* (Singapore: World Scientific, 2013), 15.

24. George Gamow to Arno Penzias, September 29, c. 1965, Arno Penzias Papers, Library of Congress, by permission of the Gamow Estate. Gamow wrote "1963," which was likely a mistake because that would place it before the CMBR discovery.

25. P. James E. Peebles, phone interview by the author, August 26, 2019.

26. Ralph Alpher, interview by Martin Harwit, August 12, 1983, American Institute of Physics, Niels Bohr Library and Archives, Oral Histories, https://www.aip.org/history-programs/niels-bohr-library/oral-histories/3014-2.

27. Ralph Alpher, interview by Martin Harwit, August 12, 1983, https://www.aip.org/history-programs/niels-bohr-library/oral-histories/3014-2.

28. R. Igor Gamow, phone interview by the author, September 2, 2019.

29. Peebles, Page, and Partridge, *Finding the Big Bang*, 374.

30. R. Igor Gamow, phone interview by the author, September 2, 2019.

31. George Gamow to Ralph Alpher, December 2, 1967, Gamow Papers, Library of Congress, by permission of the Gamow Estate.

32. Freeman Dyson, personal communication with the author, February 22, 2019.

33. George Gamow to Paul Dirac, August 17, 1968, Gamow Papers, Library of Congress, by permission of the Gamow Estate.

34. F. Hoyle and J. Narlikar, "Conformal Invariance in Physics and Cosmology," in *Cosmology, Fusion & Other Matters: George Gamow Memorial Volume*, ed. Frederick Reines (Boulder: Colorado Associated University Press, 1972), 15–28.

35. "Ralph A. Alpher: National Medal of Science, Physical Sciences, 2005," National Science and Technology Medals Foundation, https://www.nationalmedals.org/laureates/ralph-a-alpher#.

CHAPTER NINE: LIFE ON THE FRINGES

1. Donald Clayton, personal communication with the author, August 11, 2019.

2. Elizabeth Jeanne Hoyle Butler, personal communication with the author, September 5, 2019.

3. Geoffrey Hoyle, personal communication with the author, September 30, 2019.

4. Donald Clayton, personal communication with the author, August 11, 2019.

5. Stan Woosley, personal communication with the author, August 25, 2019.

6. Robert V. Wagoner, personal communication with the author, November 1, 2019.

7. Simon Mitton, *Fred Hoyle: A Life in Science* (New York: Cambridge University Press, 2011), 20.

8. Mitton, *Fred Hoyle*, 287.

9. "Take the World from Another Point of View" (interview with Richard Feynman), Yorkshire Television, 1973, http://calteches.library.caltech.edu /35/2/PointofView.htm.

10. Jocelyn Bell Burnell, interview by David DeVorkin, May 21, 2000, American Institute of Physics, Niels Bohr Library and Archives, Oral Histories, https://www.aip.org/history-programs/niels-bohr-library/oral-histories/31792.

11. Jocelyn Bell Burnell, "Jocelyn Bell Burnell at Perimeter: A Special Public Lecture Webcast," October 25, 2018, https://insidetheperimeter.ca /jocelyn-bell-burnell-perimeter-webcast/.

12. Donald Clayton, personal communication with the author, August 11, 2019.

13. Geoffrey Hoyle, personal communication with the author, September 30, 2019.

14. Nobel Prize Nomination Archive, The Nobel Prize, https://www .nobelprize.org/nomination/archive/manual.html.

15. Sarah L. Burbidge, phone interview by the author, November 3, 2018.

16. Jayant Narlikar, personal communication with the author, August 2, 2019.

17. Stan Woosley, personal communication with the author, August 25, 2019.

18. Elizabeth Jeanne Hoyle Butler, personal communication with the author, September 8, 2019. She learned about the matter through Geoff Burbidge and Hoyle's publisher, Bruce Armbruster of University Science Books, who had spoken with Geoff about his written communications with Hans Bethe. Bethe's role is speculative until the nominations for the 1983 Prize are released some years from now.

19. Geoffrey Burbidge, "Hoyle's Role in B^{2}FH," *Science* 319 (March 14, 2008): 1484.

20. Geoffrey Hoyle, personal communication with the author, September 30, 2019.

21. Robert V. Wagoner, personal communication with the author, November 1, 2019.

22. David Bergamini, "The Big Space-Fossil Mystery," *Life*, December 8, 1961, 45.

23. Chandra Wickramasinghe, *A Journey with Fred Hoyle* (Singapore: World Scientific, 2013), 147.

24. Wickramasinghe, *A Journey with Fred Hoyle*, 147.

25. Wickramasinghe, *A Journey with Fred Hoyle*, 154.

26. Wickramasinghe, *A Journey with Fred Hoyle*, 174.

27. Idan Ginsburg, Manasvi Lingam, and Abraham Loeb, "Galactic Panspermia" (draft manuscript submitted to *Astrophysical Journal Letters*), https://arxiv.org/pdf/1810.04307.pdf.

28. Ralph Alpher, interview by Martin Harwit, August 12, 1983, American Institute of Physics, Niels Bohr Library and Archives, Oral Histories, https://www.aip.org/history-programs/niels-bohr-library/oral-histories/3014-2.

29. Stan Woosley, personal communication with the author, August 25, 2019.

30. P. James E. Peebles, *Cosmology's Century: An Inside History of Our Modern Understanding of the Universe* (Princeton, NJ: Princeton University Press, 2020), 167.

31. Simon Mitton, *Fred Hoyle: A Life in Science* (New York: Cambridge University Press, 2011), 317.

CONCLUSION: THE LEGACIES OF GAMOW AND HOYLE

1. Viviana Mossa, K. Stöckel, F. Cavanna, F. Ferraro, M. Aliotta, F. Barile, D. Bemmerer, A. Best, A. Boeltzig, C. Broggini, et al. "The Baryon Density of the Universe from an Improved Rate of Deuterium Burning," *Nature* 587 (2020): 210–213.

2. Elizabeth Jeanne Hoyle Butler, personal communication with the author, September 5, 2019.

3. Stan Woosley, personal communication with the author, August 25, 2019.

4. John Osmundsen, "A Little Science for Everyone: Selected Softcover," *New York Times*, June 12, 1960.

5. Fred Hoyle, *October the First Is Too Late* (London: Heinemann, 1966), 67.

6. Robert W. Wilson, phone interview by the author, October 17, 2019.

7. P. James E. Peebles, "How Physical Cosmology Grew," Nobel Lecture, December 8, 2019, https://www.nobelprize.org/prizes/physics/2019/peebles/lecture/.

8. Jim Peebles, "Have We Got the Universe Right?" *New Scientist*, June 6, 2020, 34.

Index

NOTE: FH stands for Fred Hoyle; GG stands for George Gamow

Paul Halpern is a professor of physics at the University of the Sciences in Philadelphia and the author of seventeen popular science books, most recently *Synchronicity*, *The Quantum Labyrinth*, and *Einstein's Dice and Schrödinger's Cat*. He lives near Philadelphia, Pennsylvania.